JOURNAL OF CYBER
SECURITY AND MOBILITY

Volume 3, No. 2 (April 2014)

Special Issue on
Next Generation Mobility Network Security

Guest Editor:
Roger Piqueras Jover

JOURNAL OF CYBER SECURITY AND MOBILITY

Aim
Journal of Cyber Security and Mobility provides an in-depth and holistic view of security and solutions from practical to theoretical aspects. It covers topics that are equally valuable for practitioners as well as those new in the field.

Scope
The journal covers security issues in cyber space and solutions thereof. As cyber space has moved towards the wireless/mobile world, issues in wireless/mobile communications will also be published. The publication will take a holistic view. Some example topics are: security in mobile networks, security and mobility optimization, cyber security, cloud security, Internet of Things (IoT) and machine-to-machine technologies.

Published, sold and distributed by:
River Publishers
P.O. Box 1657
Algade 42
9000 Aalborg
Denmark

Tel.: +45369953197
www.riverpublishers.com

Journal of Cyber Security and Mobility is published four times a year.
Publication programme, 2014: Volume 3 (4 issues)

ISSN 2245-1439 (Print Version)
ISSN 2245-4578 (Online Version)
ISBN 978-87-93237-05-6 (this issue)

JOURNAL OF CYBER SECURITY AND MOBILITY COMMUNICATIONS

Volume 3, No. 2 (April 2014)

Editorial Foreword

The Long Term Evolution (LTE) is the newly adopted standard technology to offer enhanced capacity and coverage for mobility networks, providing advanced multimedia services beyond traditional voice and short messaging traffic for billions of users. This new cellular communication system is the natural evolution of 3rd Generation Partnership Project (3GPP)-based access networks, enhancing the Universal Mobile Telecommunications System (UMTS). LTE introduces a substantial redesign of the network architecture resulting in the new eUTRAN (Enhanced Universal Terrestrial Radio Access Network) and the EPC (Enhanced Packet Core). In this context, the LTE Radio Access Network (RAN) is built upon a redesigned physical layer and based on an Orthogonal Frequency Division Multiple Access (OFDMA) modulation, which features robust performance in challenging multipath environments and substantially improves spectrum efficiency and capacity. Moreover, the new all-IP core architecture is designed to be more flexible and flatter, with a logical separation of the traffic related functions and the control and signaling functions.

Over the last few years, researchers have published and presented potential vulnerabilities and attacks against the second generation of mobile networks, the Global System for Mobile Communications (GSM). Although technology, encryption schemes and security in general have drastically improved in LTE with respect of legacy GSM networks, security research should continue proactively working towards secure and resilient mobile communication systems.

In this context, the cyber-security landscape has changed drastically over the last few years. It is now characterized by large scale security threats and the advent of sophisticated intrusions. These new threats illustrate the importance of strengthening the resiliency of mobility networks against security attacks, ensuring this way full mobility network availability.

This special issue of the Journal of Cyber Security and Mobility addresses research advances in mobility security for LTE next generation mobility networks and other wireless communication systems. Two of the papers in this special issue analyze the threat of radio jamming. Lichtman et al. introduce a sophisticated reinforcement learning anti-jamming technique aimed at protecting frequency hopping wireless systems from jamming attacks while maximizing link throughput. Similarly, Kornemann et al. introduce a jamming detection mechanism for Wireless Sensor Networks (WSN), which are known to be substantially vulnerable to this type of attack.

Focusing specifically on LTE mobile networks, Jermyn et al. perform a thorough simulation analysis of potential Denial of Service attacks against the eUTRAN by means of a botnet of malware-infected mobile terminals saturating the radio interface with large amounts of traffic. This type of malware infections and the connections of such a botnet to the malicious command and control servers could be detected with the system introduced by Bickford et al. This advanced detection scheme is capable to detect malicious activity in mobile terminals and proceed to mitigate and block the malicious applications.

Finally, Alzahrani et al. present a very detailed and thorough analysis of the security research published over the last few years on mobile malware analysis, detection and mitigation. The valuable insights in this paper set a common ground and framework to facilitate research in this area, as well as to share results and easily replicate experiments.

This special issue of the Journal of Cyber Security and Mobility introduces a collection of high quality papers in the area of next generation mobile network security that will continue sparking the interest in this important area of security research. I would like to thank the editorial board members, steering committee members, advisory board members, reviewers and the staff of River Publishers for their efforts in preparing the publication of this special issue of the journal. I would also like to thank the authors of the papers in this issue for their valuable contributions to the field of mobile network security.

Roger Piqueras Jover
Senior Member of Technical Staff
AT&T Security Research Center

Characterizing Evaluation Practices
of Intrusion Detection Methods
for Smartphones

Abdullah J. Alzahrani, Natalia Stakhanova, Hugo Gonzalez
and Ali A. Ghorbani

*Information Security Center of Excellence, Faculty of Computer Science,
University of New Brunswick*
{a.alzahrani, natalia, hugo.gonzalez, ghorbani}@unb.ca

Received 28 February 2014; Accepted 15 April 2014;
Publication 2 July 2014

Abstract

The appearance of a new Android platform and its popularity has resulted
in a sharp rise in the number of reported vulnerabilities and consequently in
the number of mobile threats. Mobile malware, a dominant threat for modern
mobile devices, was almost non-existent before the official release of the
Android platform in 2008. The rapid development of mobile platform apps and
app markets coupled with the open nature of the Android platform triggered an
explosive growth of specialized malware and subsequent search for effective
defence mechanisms. In spite of considerable research efforts in this area, the
majority of the proposed solutions have seen limited success, which has been
attributed in the research community to the lack of proper datasets, lack of
validation and other deficiencies of the experiments. We feel that many of
these shortcomings are due to immaturity of the field and a lack of established and organized practice. To remedy the problem, we investigated the
employed experimentation practices adopted by the smartphone security community through a review of 120 studies published during the period between
2008–2013. In this paper, we give an overview of the research in the field
of intrusion detection techniques for the Android platform and explore the
deficiencies of the existing experimentation practices. Based on our analysis

Journal of Cyber Security, Vol. 3 No. 2, 89–132.
doi: 10.13052/jcsm2245-1439.321

we present a set of guidelines that could help researchers to avoid common pitfalls and improve the quality of their work.

Keywords: intrusion detection, smartphones, mobile malware.

1 Introduction

The rapid evolution of various mobile platforms in the past decade has swiftly brought smartphones to all aspects of our daily life. Such popularity has stimulated underground communities, giving an unprecedented rise to mobile malware. Among the most targeted platforms is the Android platform, mostly due to the ease of use of malicious apps, and the lack of proper defence. According to Kaspersky's estimation, the number of mobile malware targeting the Android platform tripled in 2012, reaching 99% of all mobile malware [37]. Also, they said that in 2013 there are more than 148,427 mobile malware modifications in 777 families and 98.05% of mobile malware found this year targets Android platform [41].

The lack of necessary defence mechanisms for mobile devices has been mostly restricted by the limited understanding of these emerging mobile threats and the resource-constrained environment of smartphones. Indeed, on the one hand, the rapid growth of vulnerabilities for a new and less-studied platform, coupled with the lack of timely access to emergent mobile malware, hinder our abilities to analyze these threats. On the other hand, the resource-constrained environment of smartphones, which is unable to afford computationally intensive operations, presents a significant challenge to the development of intelligent intrusion detection solutions. With mobile phone security quickly becoming an urgent necessity, researchers have started focussing their attention on the problem.

In the past several years the number of studies in the field of mobile phone security has been steadily increasing. In light of recent work around security-related studies in long established domains (e.g. anomaly detection), a lack of scientific rigor has been shown in the experimentation in the majority of these studies [42–43, 47]. As a result, we examine the evaluation practices of a newly appearing field of mobile phone security.

In this paper, we explore research in the area of intrusion detection for the mobile platform published during the period between 2008–2013. Aiming to discover the problems related to experimentation rigor encountered in other fields, we highlight the most common shortcomings and offer a set of guidelines for proper evaluation practice to the smartphone intrusion detection community. Within this study we give an overview of the common trends in smartphone intrusion detection research highlighting the general focus of the research efforts and the existing gaps. We hope that these efforts will give an insight into the future development of viable intrusion detection mechanisms for mobile devices.

The rest of the paper is organized as follows: in Section 2, we present some related works; in Section 3, we discus intrusion detection in mobile devices; in Section 4, we provide our assessment methodology; in Sections 5–6, we discuss the results of our evaluation; in Section 7, we present a set of guidelines that would help researchers to avoid common pitfalls and improve the quality of their work. Finally, we summarize our conclusion in Section 8.

2 Related Work

With the recent burst of research interest in the area of smartphone security, a number of studies have been aiming to organize and classify the existing efforts. One of the first attempts to summarize the research in the area of security for mobile devices was presented by Enck [24]. A broader study focusing on a variety of mobile technologies (e.g., GSM, Bluetooth), their vulnerabilities, attacks, and the corresponding detection approaches, was conducted by Polla et al. [33]. A more thorough analysis of research in the area of smartphone related to security solutions was offered by Shahzad [46]. Before these major surveys there were several other studies focusing on various aspects of mobile phone security [49, 19, 30, 12].

The lack of clear guidelines for structuring and analyzing existing research in the area of smartphone security has triggered additional efforts aiming to devise a structured taxonomy and provide necessary classification criteria. Among these efforts, there are a taxonomy for classification of smartphone malware detection techniques proposed by Amamra et al. [12], and a classification of attack vectors for smartphones developed by Becher [17].

To complement these research efforts, several study groups have been surveying mobile malware characteristics. Felt et al. [27] evaluated the behavior of 46 mobile malware samples and the effectiveness of existing defence mechanisms. On a broader scope, Zhou et al. [52] gave a detailed characterization of over 1000 Android malware samples.

This paper, on the other hand, steps beyond traditional survey boundaries and takes a critical look at the experimentation practices adopted by the smartphone security community.

3 Specificity of Intrusion Detection in Mobile Devices

Intrusion detection in traditional networks is one of the most well defined and extensively studied fields. Intrusion detection in mobile networks generally falls under an umbrella of this broader domain, and thus the core foundation of intrusion detection generally follows the defined principles. At the same time, there are several specificities that make traditional IDSs not suitable for mobile devices:

- *Constrained resources*: the resource-constrained environment of smart-phones puts strict limitations on the usage on the available time, power, and memory resources, essentially dictating what actions the detection system can and cannot afford. As such many of the approaches that require a heavy computational operations (e.g., malware static analysis) are avoided.
- *Mobility*: As opposed to traditional IDSs where an IDS system is perma-nently stationed on a known network or a host, mobile device IDSs are generally located on a mobile device with some more resource intensive functionality residing on a cloud. Thus as mobile device goes through a variety of networks with often unknown configurations and different security postures, mobile IDS faces various challenges to provide a comprehensive defense for a wide range of threats and conditions.
- *Deployment environment*: One of the security features characterizing modern mobile platforms is the use of sandbox environment that allows to constrain unwanted activity. Since a sandbox is meant to execute untrusted code, trusted native code is generally run directly on a platform. Although sandboxing is generally seen as a desirable intrusion detection technique, it has limitations. Sandboxing is usually less effective in detecting non-generic targeted attacks, e.g., malware designed to be activated on specific user action or to trigger malicious behavior after a period of normal activity.

  Sandboxing is also largely ineffective against another practice, i.e., the use of external code, that have been gaining popularity in mobile apps. This mechanism allows to use legitimate application to load a malicious functionality without requiring any modifications to the existing legit-imate code. As such the original bytecode remains intact allowing an app evade detection. Poeplau et al. [40] defined several techniques to load external code on a device: with the help of class loaders that allow to extract classes from files in arbitrary locations, through the package context that allow to access resources of other apps, through the use of native code, with the help of runtime method exec that gives access to a system shell, and through less stealthy installation of .apk files requested by a main .apk file.

- *Exposure*: The typical attacks vectors seen in traditional platforms also exist in mobile environments. However, the specificity of mobile devices opened up new avenues for compromise. As such infection methods usually not monitored by traditional IDS, e.g., through SMS, MMS, app' markets, have recently gained a wide popularity.
- *Privacy*: As opposed to traditional networks where privacy leakage typically constitutes a small portions of potential threats, private information theft is rapidly becoming one of the major concerns for mobile devices [32, 44].

4 Evaluation Methodology

To investigate evaluation practices employed by security community, we conducted a survey of research work in the area of intrusion detection for smartphones published since the official release of Android platform in 2008. To avoid selection bias, we collected all research papers indexed by Google Scholar for the reviewed time period from 2008 until 2013. This included studies introducing defence mechanisms specifically developed for the smartphone platforms.

Our research study excluded short papers, extended abstracts, non-peer-reviewed research, and papers not available in the English language. To narrow our focus, we further selected research work relevant to intrusion detection; thus, any methods specifically developed for fault detection, safety analysis, etc. were excluded. The final set of 120 papers, containing 17 journal and 103 conference/workshop papers, was reviewed manually without use of any automatic search techniques. Each of the selected papers were put through at least two evaluation rounds to reduce classification errors.

5 Overview of the Reviewed Studies

In the past several years the number of studies in the field of mobile phone security has been steadily increasing as our survey shows in Figures 1, 2.

Even though most of the research focus on Android, we have seen other platforms used as testing platform as illustrated in Table 1. Traditionally, there have been various classifications relating to intrusion detection mechanisms. Statistics about the surveyed papers with regards to these classifications are

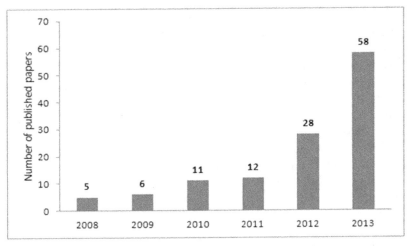

Figure 1   The reviewed papers: a perspective over the years

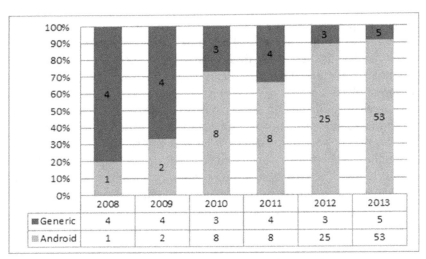

Figure 2   The reviewed papers: a perspective over the years

Table 1   The details of generic approaches

Testing Platforms	
Symbian OS	35% (8 papers out of 23)
Windows Mobile OS	26% (6 papers out of 23)
Hybrid	39% (9 papers out of 23)

Table 2 The details of the surveyed papers

Papers by intrusion detection types	
Network-Based methods	5% (6 papers out of 120)
Hybrid Methods	13% (15 papers out of 120)
Host-Based Methods	82% (99 papers out of 120)
Application level	64% (73 papers out of 114*)
Operating system level	3% (4 papers out of 114)
Hardware level	4% (5 papers out of 114)
Hybrid	29% (32 papers out of 114)
Malware Detection	67% (80 papers out of 120)
By detection approach:	
Anomaly-Based	55% (44 papers out of 80)
Signature-Based	44% (35 papers out of 80)
Hybrid	1% (1 papers out of 80)
By focus:	
Malicious Apps	65% (52 papers out of 80)
Information Leakage	19% (15 papers out of 80)
System Behavior	16% (13 papers out of 80)
Papers by applied detection approach	
Anomaly-Based	58% (70 papers out of 120)
Signature-Based	40% (48 papers out of 120)
Hybrid	2% (2 papers out of 120)
Papers by a level of invasiveness	
Static	35% (42 papers out of 120)
Dynamic	48% (57 papers out of 120)
Hybrid	17% (21 papers out of 120)

shown in Table 2. Among the reviewed papers, the majority of the studies focused on a host-based detection (99 papers out of 120), with only six papers introducing network-based mechanisms and 15 papers proposing hybrid approaches. In addition to these common categories, we noticed an increased interest in specialized mechanisms for mobile malware detection: 80 papers out of 120 considered various aspects of malware detection either through detection of malicious apps (52 out of 80 papers), detection of information leakage (15 out of 80 papers) or suspicious system behavior (13 out of 80 papers).

5.1 Intrusion Detection Focus

To provide a broad overview of research conducted in the field, we categorize the research based on three common parameters: monitoring scope of the proposed technique, underlying detection approach and level of

---

*114 papers include 99 pure host-based analysis and additional 15 hybrid approaches.

technique invasiveness. Figure 3 illustrates a high-level summary of these categorizations.

By scope of monitoring In general, suspicious activity can be spotted at least at one of the following levels: application, operating system (OS) or hardware. Events at the application level are generally related to user activity and are often seen through suspicious SMS/MMS messages, unusual keystroke dynamics, etc. Since this is a high-level behavior that is not necessarily exhibited by all malware threats, a placement of detection mechanisms at this level provides only limited coverage. OS level activity, on the other hand, includes events triggered by the built-in OS mechanisms (e.g., system calls) thus giving a better picture of the underlying system behavior. While this is often a preferred location for intrusion detection mechanisms in traditional computer-based environments, it presents a number of problems for resource-constrained mobile phone environments that can only afford lightweight detection techniques. Finally, intrusion detection at hardware level allows researchers to obtain basic measurements of the monitored systems (e.g., CPU, network usage, power supply) that might be indicative of abnormal device behavior, especially when they are compared to normal device usage.

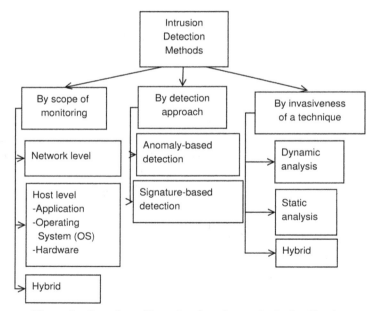

Figure 3   Overview of intrusion detection methods classification

Detection at this level provides a number of benefits that are particularly valuable for mobile devices, including fast, reliable, and scalable intrusion detection. Of the reviewed papers, the majority of studies focus on application level (64%), then OS level (only 3%), and then hardware level (4%). The remaining papers address hybrid levels (29%).

By detection approach Another common classification of intrusion detection techniques, based on how intrusive activities are detected, broadly groups techniques into anomaly-based and signature-based approaches. Since anomaly-detection is perceived as more powerful due to its higher potential to address new threats, it is not surprising to see the majority of the reviewed studies employing it (70 papers out of 120).

Interestingly, a large portion of the reviewed studies is focused on the application of signature-based detection. In spite of criticism of this approach in academic research (due to a high reliance on ever-growing signature bases and the lack of detection of 'zero-day' threats) almost 40% of the reviewed studies (48 out of 120) employed signature-based detection.

By invasiveness of technique The approaches for detection of intrusive activity can be further broken down into dynamic (behavior-based) and static detection, depending on the level of invasiveness of the intrusion detection system. Dynamic techniques generally adopt a black box approach by focusing on behavior of a target system in general or specific files/apps in particular. Since this detection is only possible during the execution of a program, it has a limited focus and is often at the mercy of the executable malware that might or might not exhibit suspicious behavior in a given running environment. Static detection, on the other hand, allows the researcher to analyze an app/file without its execution (via disassembly or source code analysis) and is thus considered to be more thorough, although it is more expensive. Among the reviewed studies, 48% of the papers employed dynamic analysis, with 35% focusing on static detection.

5.2 Trends

The vast popularity of mobile phones, and specifically the Android platform resulted in a rapid increase of mobile malware. One of the major sources of malware became third party app markets [29, 52]. For example, due to popularity of Android platform, we have seen a arise of alternative Android app markets (i.e. not officially supported by Google), often known for their lack of security checks and thus favored by malware writers. Figure 4 clearly shows this trend. Although the use of Google Play market slightly increased

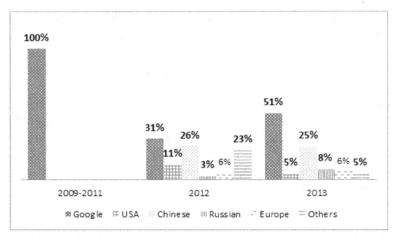

Figure 4   The use of apps from various markets in the surveyed works

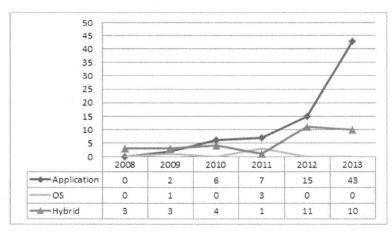

Figure 5   Trends of Research Focused

from 2012 to 2013, it has a significantly smaller share compared to the newly-emerging markets.

Among other trends noticed in the reviewed studies is a clear tendency towards application-level detection (Figure 5). The interest in application-level detection has been steadily increasing and in 2013 the proportion of studies with this scope reached 64%.

6 Detailed Survey Observations

The impact of data quality on the outcomes of the various studies was emphasized throughout the research. Indeed, the ability of a study to predict the performance of a method in a real deployment environment is highly dependent on the ability of a dataset to reflect conditions of that deployment setting.

Datasets. Although the mobile phone setting is not an exception, the lack of comprehensive datasets (mostly attributed to immaturity of the field) has posed a significant problem [52]. There have been several attempts to create structured, comprehensive datasets specifically for evaluating mobile security solutions. Table 3 lists some of these publicly available resources. In spite of their variety, most of these sets have limited applicability and/or require additional preprocessing. The lack of appropriate datasets is reiterated throughout the papers in our survey: all papers that involved experimental studies (110 out of 120 papers) employed self-constructed datasets (see table 4). Although the researchers' intentions are understandable, the lack of standardized datasets employed across the majority of studies raises several concerns.

The first concern is the transparency of a dataset. Customized datasets require collecting data, preprocessing activities (e.g., anonymization, cleaning out redundant data, assuring necessary balance of normal and suspicious behavior), and validating (e.g., confirmation of 'ground truth'). These activities are tedious, time-consuming and often require specialized infrastructure, including equipment and necessary permissions for collecting private data. Unless these activities are fully described and the dataset made available, there is little assurance that a dataset will be representative of a deployment environment. Although the initial step can be avoided with the use of malware repositories, the rest of the concerns remain.

Due to the nature of smartphone threats, the authors of all 110 papers that conducted an experimental study employed a set of mobile apps for this analysis. Out of these 110 papers, 91 papers used real apps collected through app markets and online mobile repositories, three papers implemented their own version of apps and five papers used a combination of these two categories (real and self-written apps).

The majority of the real apps used in surveyed papers were collected through various Android markets (58% of papers) with the assumption that apps coming from a market are benign. With the increasing number of reports showing markets being adopted by hackers as one of the ways for malware distribution, it is expected that all market apps in any experimentation study

Table 3 The existing sources of data for evaluation of mobile security solutions

Name	Institution	Description	Access	Size	Fixed	Labeled	Year	Platform
MIT Reality mining dataset [23]	MIT Human Dynamics Lab	Data collected on smartphones on users' location, communication and device usage behavior for over 9 months; includes only normal user behavior.	Free registration	–	Yes	Yes	2005	Collected on Nokia platform
Android Malware Genome Project [52]	North Carolina State University	Collected malware samples from 49 representative Android families.	Academic registration	1260 samples	Yes	Yes	2011	Android
Mobile Malware in the wild dataset [27]	University of California, Berkeley	Collected malware in the wild from 2009 to 2011	Public Information, not samples.	46 samples	Yes	Yes	2011	iOS, Symbian, Android
Contagio mobile [4]	Personal	Digital place to share mobile malware samples and Analysis, based on a centralized community effort.	Public	330 samples	No	Yes	2011	All mobile
Google group mobile. malware [8]	Community effort	A mailing list for researchers in mobile malware field. Contains material related to new mobile malware samples, analysis, new techniques, questions pertaining to the field, and other related data.	Free registration	–	No	Partially	2010	iOS, Android, Symbian, Windows mobile

Name	Source	Description	Access	Size			Year	Platform
AndroMalShare [1]	Xi'an Jiaotong University	This project focuses on sharing Android malware samples. A sampled scanned by SandDrois and provides a detailed report and detection results from several anti-virus tools.	Academic registration	8745 samples	No	Yes	2013	Android
VirusShare [10]	Personal	A personal effort to share samples collected over the years, and accepting contributions from others.	By invitation	–	No	Partially	2011	All type, including mobile
VirusTotal Malware Intelligence Services [11]	Google subsidiary	Free online service for analysis of suspicious binaries and URLs.	Public, only analysis	–	No	Yes	2002	All type, including mobile
SMS Spam Collection [9]	Personal	Collection of SMS messages used for mobile phone spam studies collected by different research groups and by crawling the Internet.	Public	5,574 messages	Yes	Yes	2006–2007	Mobile phones
Android Malware Performance Counter Data [21]	CASTL, Columbia University	Collection of performance counters for Android.	Free registration	–	Yes	Yes	2012	Android ARM

Table 3 Continued

Name	Institution	Description	Access	Size	Fixed	Labeled	Year	Platform
Mining Permission Request Patterns [28]	University of California, Berkeley	The dataset provides the permissions requested by the application., the price, the number of downloads, the average user rating, and a short prosaic description	Public	188,389	Yes	Yes	2011	Android
DroidAnalytics [51]	ANSRLab, Chinese University of Hong Kong	Android malware from 102 different families, with 342 of them being zero-day malware samples from six different families.	Academic registration	2494	Yes	Yes	2013	Android
Drebin [15]	University of Gottingen	Android malware from 176 different families.	Academic registration	5560	Yes	Yes	2010–2012	Android

Table 4 Experimentation practices employed in the surveyed works

Datasets	
Data collection detail:	
Market-Based	52% (63 papers out of 120)
Host-Based	43% (51 papers out of 120)
Network-Based	5% (6 papers out of 120)
Datasets details:	
Self-constructed set	97%(107 papers out of 110)
MIT Reality	3%(3 papers out of 110)
Dataset sharing	3% (3 papers out of 110)
Real apps	83% (91 papers out of 110)
Self-written	11% (12 papers out of 110)
Hybrid	6% (7 papers out of 110)
Normal apps:	
From Market	54% (53 papers out of 98)
Not specified	46% (45 papers out of 98)
Malware samples:	
From Repositories	16% (16 papers out of 98)
From Known Data set	17% (17 papers out of 98)
Hybrid	2% (2 papers out of 98)
Not specified	65% (63 papers out of 98)
Evaluation	
Performed experimental study	90% (108 papers out of 120)
Used simulation	2% (2 papers out of 120)
Do not perform experimental study	8% (10 papers out of 120)
Among 110 that performed experiments:	
Reported evaluation results	93% (102 papers out of 110)
Compared with other techniques	15% (17 papers out of 102)
Involved Internet Access	15% (31 papers out of 78*)
Used Monkey tool	15% (11 papers out of 78)
Employed evaluation metrics:	
Detection Rate	23% (25 papers out of 110)
FPR	35% (39 papers out of 110)
Recall	26% (29 papers out of 110)
Precision	10% (11 papers out of 110)
ROC curves	11% (12 papers out of 110)
AUC	12% (13 papers out of 110)
Self-developed metrics	11% (12 papers out of 110)
DR or FPR or ROC Curves	45% (50 papers out of 110)

will undergo a thorough check to ensure their legitimacy. However, among these studies only 27 verified that apps are malware free.

---

*78 papers include 57 dynamic analysis and additional 21 hybrid methods.

The other concern related to the use of several sources (i.e., markets) for compiling a single dataset is the necessity of data cleaning. Since many authors distribute apps through a number of markets, duplicative apps (both legitimate and malware) are commonly encountered in different markets. In the reviewed set only 27% of the papers reported at least some activity related to data cleaning.

Although malware apps can be obtained from a number of sources, only 16 out of 98[1] papers utilized existing mobile malware repositories, while several papers reported the use of self-implemented malware apps. With the appearance of research-based mobile datasets, several studies have turned their attention to already prepared data. As such, 17 out of 98 papers used samples from existing malware sets. However, all of these studies have only partially used the datasets, either removing or complementing the existing samples with additional data. This move is another indication of a dire lack of suitable datasets in mobile phone security.

Table 5 presents a summary of the distribution of legitimate and malicious apps in the employed datasets. Throughout our analysis, we observed a very large differences in the sizes of employed datasets. The main concern that this variability raises is the presence of studies with a very small number of samples. Given the availability of data, especially in recent years, evaluations performed on only three malware samples is hardly representative and thus unjustifiable.

The second concern relates to the lack of standardized datasets available to researchers. Among the reviewed studies only three papers made datasets publicly available[2].

Table 5   The sizes of the employed data set in the surveyed papers

Years	Total Size		Normal Samples		Malware Samples	
	Smallest	Largest	Smallest	Largest	Smallest	Largest
2013	8	276016	3	150368	3	12158
2012	6	482514	20	207865	5	378
2011	5	42000	2	13098	3	32
2010	1	2285	30	2285	1	5
2009	240	311	311	311	2	240
2008	4	1000	3	1000	4	7
All years	1	482514	3	207865	1	12158

[1]The 98 papers include 91 studies employing real apps and seven studies using hybrid datasets.

[2]An explicit note about this availability was made in the content of the paper.

The third concern relates to the feasibility of the comparative analysis. Such variability of customized data makes comparative analysis of techniques challenging. Indeed, among the surveyed papers only 15% (17 out of 102 papers) attempted to compare the effectiveness of the developed approach with other methods.

The fourth concern and one of the primary ones when selecting suitable datasets for experiment, is the selection of features that would serve as a basis for analysis. During our survey we extracted 188 unique features used in the reviewed papers. All the papers, even the ones that did not involve the performance of experiments, propose features for analysis. These features were classified according to the categorizations outlined in Figure 6. The statistics for the most commonly employed features are given in Table 6. In spite of 'permissions' being the most commonly used feature throughout the last five years (29% of papers), before that (prior to 2011) 'system calls' was the most popular feature (as illustrated in nine papers out of 22 papers published during 2008–2010). The variability of features in the surveyed research is yet another indication of the immaturity of the field.

Experimentation. Evaluation of the proposed approach is an essential component of any research study in the domain of intrusion detection. Among the reviewed papers, 10 (8%) did not involve any experimentation, while one paper limited its analysis to simulation only. The limitations of simulation in security domain have been repeatedly emphasized in academic studies [13, 39]. Seen as non-exhaustive evaluation technique, simulation does not provide necessary depth of the analysis mostly due to inability to

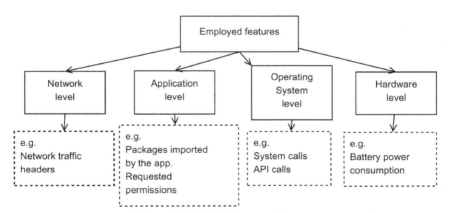

Figure 6   A classification of features

Table 6   Features employed in the surveyed papers

Application level:		
Requested permissions	29%	(69 papers out of 120)
Imported Package	3%	(8 papers out of 120)
Extra information	6%	(14 papers out of 120)
Operating System level:		
API calls	16%	(38 papers out of 120)
System calls	19%	(31 papers out of 120)
Control flow graphs	1%	(3 papers out of 120)
Dataflow	1%	(2 papers out of 120)
Logged Behavior sequences	8%	(19 papers out of 120)
Instructions (Opcode)	6%	(14 papers out of 120)
Hardware level:		
Power consumption	6%	(17 papers out of 120)
Network level level:		
Network traffic	11%	(27 papers out of 120)

guarantee security properties. Constrained to a given scenario simulation does not give insight into a method's performance in unrestricted threat environments or with undiscovered attacks. Although a thorough simulation can provide a sense of average performance of the evaluated technique, it should not be utilized for comprehensive evaluation of security properties.

Among the studies that lacked experiments, only one work reported a proof-of-concept implementation with no mentioning of results. Two of these studies gave a theoretical analysis of the introduced method's performance. Close analysis of the rest of the studies revealed that while all of them discussed some strategies for implementation and potential analysis, none of them offered neither of these.

The details of experimental setup and the employed methodology are necessary to ensure repeatability of the experiments and therefore to facilitate the comparison between techniques. However, as our survey shows most of the researchers neglect to include these details in the study description.

For example, user interface interactions, as one of the triggers of malicious behavior, are known to be essential for analysis of mobile malware [50]. However, among 78 papers that performed dynamic analysis and hybrid analysis of malware, only 11 papers reported the use of user interface interactions to produce events. All of these studies however employed the use of Monkey tool that produces pseudo-random streams of user and system events [14]. Meant for general stress testing of Android apps, the tool is limited to some subset

Table 7   The details of published techniques

Compared with other techniques	
Virus total scanner [11]	23% (4 papers out of 17)
TaintDroid [25]	17% (3 papers out of 17)
Kirin [26]	12% (2 papers out of 17)
Andromaly [45]	12% (2 papers out of 17)
VirusMeter [36]	6% (1 papers out of 17)
Andrubis [35]	6% (1 papers out of 17)
DNADroid [20]	6% (1 papers out of 17)
Androguard [22]	6% (1 papers out of 17)
Clam-AV Scanner [31]	6% (1 papers out of 17)
Manual Analysis	6% (1 papers out of 17)

of actions that do not correctly represent a realistic user of system behavioral patterns.

Similarly, although the access to the Internet is one of the important triggers for many malware, we found that only 31 out of 78 studies reported the use of Internet access in their experiments.

Evaluations. Among the 110 papers that undertook evaluation, eight studies did not report results. Of the 110 studies overall, most works (102) offered experimentation results, which varied significantly from extensive discussion reasoning behind the obtained numbers to a brief mentioning of whether an attack was detected or not.

The evaluation of the method's effectiveness is generally performed along two directions: (1) evaluation of method's overhead in terms of CPU occupancy, memory usage, power consumption and detection speed; and (2) evaluation of the method's detection capability. While both types of experiments are necessary to assess the quality of the newly developed approach, the majority of the surveyed studies are mostly focused on the evaluation of classification accuracy of a method. To assess this accuracy though, it is necessary to compare the new method's performance against known benchmarks in the field. However, due to the infancy of the field such benchmarks are mostly non-existent. As such most of the studies either do not provide any comparison or look at previously proposed methods. Of these surveyed papers, only 17 (15%) compared experimentation results with perviously published intrusion detection techniques and tools. The list of these techniques is given in Table 7. The fact that these techniques are selected for comparison might be an indication of them becoming future benchmarks.

7 Guidelines

The evaluation of intrusion detection methods is an inherently difficult problem due to a variety of factors, from a diversity of monitored events to the unpredictable nature of a deployed environment. Aggravated by the immaturity of the field, evaluation of intrusion detection solutions for smartphone platforms are often guided by subjective choices and individual preferences of the researchers. A review of the experimentation practices employed in this field suggests several important points. Based on the results of our review we formulated several guidelines for proper evaluation of intrusion detection methods for smartphones.

The resulting criteria is offered in three general dimensions of scientific experimentation, formulated by previous research [47, 43]: factors related to *the employed datasets*, *the performed experiments*, and *the performance evaluation*. Since the previous studies have attempted to outline some of the limitations and constraints of scientific experimentation in computer security, we aimed to devise comprehensive guidelines and recommendations for a smartphone setting. The following guidelines can be used as a structure for experimental design and subsequently in manuscript organization:

Employed datasets define applicability and deployability of the developed technique to a real-life setting. To ensure the effectiveness of these qualities the dataset description is an essential component of any manuscript. Although this description will vary depending on the source of data (self-constructed or publicly available), it should provide enough detail to allow an intelligent analysis of the proposed technique. Specifically, the following aspects of employed data should be addressed:

- Data overview:
  - the source of the data, (i.e., whether the dataset is public, proprietary, or self-constructed).
  - quantitative description of malicious and normal events in a dataset (e.g., apps, malware, network flows).
  - how these malicious and normal events were obtained (i.e., simulated, implemented, collected). This should include description of the environment/process. For example, assuming the dataset includes Android apps that are collected from various sources: are these apps free or paid, what are the sources (e.g., list app markets), what malware families these apps represent, why these malware apps are selected while others are excluded? On the other hand, if

the data is collected on the network, then the description should include the duration/time of the collection, access to the Internet, use of visualization, etc.

- monitored features.
- Data validation:
  - the ground truth established, (i.e., how the legitimacy and abnormality of data is verified). For example, in the case of Android apps, whether or not the collected apps checked through third-party sources.
  - obsolete data removed, (i.e., in the case where a dataset containing malware or apps is employed, sinkholed or inactive samples should be removed).
- Training/testing data:
  - quantitative division of dataset into training and testing sets, corresponding numerical estimation of normal and malicious samples in each.
- Data sharing:
  - data employed in the experiments should be archived for reference for future authors. Even if datasets are not made public, the possibility of sharing on demand should be explicitly indicated in a manuscript.

The performed experiments refer to the setting of the performed experiments and thus allow for transparency of the experiments to be ensured, (i.e., that the conducted experiments are repeatable and comprehensible). Several aspects need to be addressed here to allow for objective evaluation of a study:

- Environmental setting:
  - experimental setup, (i.e., simulation, emulation, experimentation).
  - context for execution for both victim and attack side, including hardware (e.g., devices employed, their topology) and software (e.g., operating systems, NAT, privileges). This description might be complemented with a diagram to avoid ambiguities in interpreting an environmental setup.
  - the employed tools, with an indication of their releases and versions, and parameters.
  - the use of technology (e.g., access to the Internet).

- Experiments:
  - methodology, (i.e., steps involved, allowed user interaction, duration, number of repetitions).
  - the use of sampling, the employed algorithm and justification.

Performance evaluation. The primary goal of this section is to give insight into a proposed method for performance, which is called for a detailed and, more importantly, objective analysis of evaluation.

- Evaluation Methodology:
  - the scripts or procedures used for analysis should be specified. Often a closer analysis of intrusion detection results call for a manual examination of traces. Regardless of the findings this should be stated.
  - the employed statistical apparatus.
- Evaluation metrics:
  - define evaluation metrics. To avoid ambiguities in metric interpretation, a clear definition of a chosen metric should be provided. This becomes critically important when non-traditional metrics are employed, as the lack of consistent terminology hinders researcher's ability to properly identify and apply common methods. As the use of self-developed metrics becomes more widespread, a metric definition should be followed by a clear validation of the proposed metric for a given task.
  - ensure a proper combination of metrics. Detection rate (DR) and false positive rate (FPR) (or their graphical representation ROC curve) are the most widely used metrics in spite of criticism. Several studies have shown that the ROC curve alone or DR/FPR metrics combination might be misleading or simply incomplete for understanding the strengths and weaknesses of intrusion detection systems [18, 48, 38, 16, 34]. Although such misinterpretation can be partially avoided when DR/FPR values are complemented with the absolute numbers of false alarms or missed detections, exclusive evaluation of an approach with these methods may not accurately represent the performance of the system.
- Findings:
  - provide numerical results. The reporting of numerical results should be comprehensive (i.e., use complementary metrics that show all

perspectives of the evaluated method performance) and consistent (i.e., reported in the same equivalents). For example, only stating the accuracy of the proposed approach (i.e., that shows the percentage of correctly classified events) gives almost no insight into a method's performance with respect to malicious events. On the other hand, for example, stating detection rates in percentages, while indicating false positive rates in absolute numbers, makes it challenging for a researcher to interpret the numbers.

- interpret the results. The intrusion detection community has traditionally focused on understanding the limitations of detection methods. Therefore, simply stating numbers is not sufficient for interpreting the performance of the detection method. Numerical results should be included to show the soundness of the approach and allow for future comparative analysis of the detection methods, but they are not the main goal. Thus, a close and often manual examination of results (e.g., false positives and false negatives) is necessary to understand and explain to the research community why and when a system performs in a given way.

- Comparison:

  - a comparative analysis of the proposed scheme with the established benchmarks in the field is an essential component of a study. Whenever existing benchmarks are not available an attempt should be made to perform a quantitative comparison against prior results.

8 Conclusion

With the popularity of Android platform, the amount of research studies on security of smartphones is rapidly increasing. While an increasing volume of studies in the field is generally seen as an indicator of a field evolution, the true value of existing work and its impact on the smartphone security progress remains unclear.

As such in spite of variability of tools proposed only a few of them have been accepted by a security community. Among them is an Android analysis tool, Androguard [22]. Effective against even obfuscation techniques, Androguard has been adopted many academic and industry malware analysis solutions, such as Virustotal portal [11], APKInspector [3], Marvinsafe [7], Anubis (Andrubis) [35], Androwarn [2], googleplay-api [5] and MalloDroid [6]. In academic studies, as our review shows, there is less consensus and studies employ a variability of tools for evaluation of new solutions.

The validity of experimental research in academic computer science in general has shown to be questionable. Through our review we also confirmed our initial hypothesis, showing that immaturity of the smartphone security negatively affects experimentation practices adopted in the field. In this context, it is plausible to suggest that the lack of adoption of the developed innovations by industry can be attributed to the lack of proper rigor in experimentation that shadows the true value of the developed solutions. Among the factors that contribute to the lack of scientific rigor in experimentation are the lack of consistency and transparency in the use of datasets, the lack of clear understanding of relevant features, the lack of benchmarked experimentation, and the biased selection of malicious apps for analysis.

While the infancy of the field of mobile phone security might justify some shortcomings in experimentation, many of the pitfalls can be avoided with proper practices established and adopted by a security community. To facilitate acceptance of this practice, we formulated a set of guidelines for a proper evaluation of intrusion detection methods for smartphone Platforms. The suggested guidelines allow a detailed description and analysis of experimentation results, and, as such, might be used for designing experimentation studies as well as for structuring experimentation sections in a manuscript. While some of the suggested guidelines might be seen as common sense, we believe that framing them in a structured way will help researchers to improve experimentation practices by providing them with a methodological reference.

9 Acknowledgment

The first author graciously acknowledges the funding from University of Hail at Saudi Arabia.

References

[1] Andromalshare. http://202.117.54.231:8080/.

[2] Androwarn. https://github.com/maaaaz/androwarn.

[3] Apkinspector. https://github.com/honeynet/apkinspector/.

[4] contagio mobile. http://contagiodump.blogspot.ca/.

[5] googleplay-api. http://www.segmentationfault.fr/publications/reversing-google-play- and-micro-protobuf-applications/.

[6] Mallodroid. http://www2.dcsec.uni-hannover.de/files/android/p50-fahl.pdf.

[7] Marvinsafe. http://www.marvinsafe.com/.

[8] Mobile malware forum. https://groups.google.com/forum/?fromgroups=#!forum/mobilemalware.

[9] Sms spam collection v.1. http://www.dt.fee.unicamp.br/tiago/smsspam collection/.

[10] Virusshare.com - because sharing is caring. http://virusshare.com/.

[11] Virustotal malware intelligence services. https://www.virustotal.com.

[12] A. Amamra, C. Talhi, and J.-M. Robert. Smartphone malware detection: From a survey towards taxonomy. In *Malicious and Unwanted Software (MALWARE), 2012 7th International Conference on*. IEEE, 2012.

[13] T. R. Andel and A. Yasinac. On the credibility of manet simulations. *Computer*, 39 (7): 48–54, July 2006.

[14] Android Developers. Ui/application exerciser monkey. http://developer .android.com/tools/help/monkey.html.

[15] D. Arp, M. Spreitzenbarth, M. Huübner, H. Gascon, K. Rieck, and C. Siemens. Drebin: Effective and explainable detection of android malware in your pocket. 2014.

[16] S. Axelsson. The base-rate fallacy and its implications for the difficulty of intrusion detection. In *CCS '99: Proceedings of the 6th ACM conference on Computer and communications security*, pages 1–7, New York, NY, USA, 1999. ACM.

[17] M. Becher, F. Freiling, J. Hoffmann, T. Holz, S. Uellenbeck, and C. Wolf. Mobile security catching up? revealing the nuts and bolts of the security of mobile devices. In *Security and Privacy (SP), 2011 IEEE Symposium on*, pages 96–111, 2011.

[18] A. A. Cárdenas, J. S. Baras, and K. Seamon. A framework for the evaluation of intrusion detection systems. In *SP '06: Proceedings of the 2006 IEEE Symposium on Security and Privacy*, pages 63–77, Washington, DC, USA, 2006. IEEE Computer Society.

[19] M. Chandramohan and H. B. K. Tan. Detection of mobile malware in the wild. *Computer*, 45(9): 65–71, 2012.

[20] J. Crussell, C. Gibler, and H. Chen. Attack of the clones: Detecting cloned applications on android markets. In *Computer Security–ESORICS 2012*, pages 37–54. Springer, 2012.

[21] J. Demme, M. Maycock, J. Schmitz, A. Tang, A. Waksman, S. Sethumadhavan, and S. Stolfo. On the feasibility of online malware detection with performance counters. In *Proceedings of the 40th annual international symposium on Computer architecture*, ISCA '13, New York, NY, USA, 2013. ACM.

[22] A. Desnos and G. Gueguen. Android: From reversing to decompilation. In *Blackhat*, 2011.

[23] N. Eagle and A. (Sandy) Pentland. Reality mining: sensing complex social systems. *Personal and Ubiquitous Computing*, 10(4): 255–268, 2006.

[24] W. Enck. Defending users against smartphone apps: techniques and future directions. In *Proceedings of the 7th international conference on Information Systems Security*, ICISS'11, pages 49–70, Berlin, Heidelberg, 2011. Springer-Verlag.

[25] W. Enck, P. Gilbert, B.-G. Chun, L. P. Cox, J. Jung, P. McDaniel, and A. Sheth. Taintdroid: An information-flow tracking system for real-time privacy monitoring on smartphones. In *OSDI*, volume 10, pages 1–6, 2010.

[26] W. Enck, M. Ongtang, and P. McDaniel. On lightweight mobile phone application certification. In *Proceedings of the 16th ACM conference on Computer and communications security*, pages 235–245. ACM, 2009.

[27] A. P. Felt, M. Finifter, E. Chin, S. Hanna, and D. Wagner. A survey of mobile malware in the wild. In *Proceedings of the 1st ACM workshop on Security and privacy in smartphones and mobile devices*, SPSM '11, New York, NY, USA, 2011. ACM.

[28] M. Frank, B. Dong, A. P. Felt, and D. Song. Mining permission request patterns from android and facebook applications. pages 870–875, 12 2012.

[29] G. Kelly. Report: 97% of mobile malware is on android. this is the easy way you stay safe. http://www.forbes.com/sites/gordonkelly/2014/03/24/report-97-of-mobile-malware-is-on-android-this-is-the-easy-way-you-stay-safe/. Accessed on Mar 2014.

[30] L. Ketari and M. A. Khanum. A review of malicious code detection techniques for mobile devices. *International Journal of Computer Theory and Engineering*, 4(2): 212–216, 2012.

[31] T. Kojm. Clamav anti-virus. http: //www.clamav.net/.

[32] B. Krishnamurthy and C. E. Wills. Privacy leakage in mobile online social networks. In *Proceedings of the 3rd Conference on Online Social Networks*, WOSN'10, pages 4–4, Berkeley, CA, USA, 2010. USENIX Association.

[33] M. La Polla, F. Martinelli, and D. Sgandurra. A survey on security for mobile devices. *Communications Surveys Tutorials, IEEE*, 15, 2013.

[34] A. Lazarevic, L. Ertoz, V. Kumar, A. Ozgur, and J. Srivastava. A comparative study of anomaly detection schemes in network intrusion detection. *Proceedings of the Third SIAM International Conference on Data Mining*, 2003.

[35] M. Lindorfer. Andrubis: A tool for analyzing unknown android applications. http://blog.iseclab.org/2012/06/04/andrubis-a-tool-for-analyzing-unknown-android-applications-2/.

[36] L. Liu, G. Yan, X. Zhang, and S. Chen. Virusmeter: Preventing your cellphone from spies. In *Recent Advances in Intrusion Detection*, pages 244–264. Springer, 2009.

[37] D. Maslennikov and Y. Namestnikov. Kaspersky security bulletin 2012. the overall statistics for 2012, 2012.

[38] J. McHugh. The 1998 Lincoln Laboratory IDS evaluation. In *RAID '00: Proceedings of the Third International Workshop on Recent Advances in Intrusion Detection*, pages 145–161, London, UK, 2000. Springer-Verlag.

[39] I. Parris, F. Ben Abdesslem, and T. Henderson. Facebook or fakebook? the effects of simulated mobile applications on simulated mobile networks. *Ad Hoc Netw.*, 12: 35–49, Jan. 2014.

[40] S. Poeplau, Y. Fratantonio, A. Bianchi, C. Kruegel, and G. Vigna. Execute this! analyzing unsafe and malicious dynamic code loading in android applications. In *Proceedings of the Network and Distributed System Security Symposium (NDSS)*, 2014.

[41] K. L. G. RESEARCH and A. TEAM. Kaspersky security bulletin 2013. the overall statistics for 2013, 2013.

[42] H. Ringberg, M. Roughan, and J. Rexford. The need for simulation in evaluating anomaly detectors. *SIGCOMM Comput. Commun. Rev.*, 38(1): 55–59, 2008.

[43] C. Rossow, C. J. Dietrich, C. Grier, C. Kreibich, V. Paxson, N. Pohlmann, H. Bos, and M. v. Steen. Prudent practices for designing malware experiments: Status quo and outlook. In *Proceedings of the 2012 IEEE Symposium on Security and Privacy*, SP '12, pages 65–79, Washington, DC, USA, 2012. IEEE Computer Society.

[44] N. Seriot. iphone privacy. In *Proceedings of the Black Hat*, Arlington, Virginia, USA, 2010.

[45] A. Shabtai, U. Kanonov, Y. Elovici, C. Glezer, and Y. Weiss. "andromaly": A behavioral malware detection framework for android devices. *J. Intell. Inf. Syst.*, 38(1): 161–190, 2012.

[46] F. Shahzad, M. A. Akbar, and M. Farooq. A survey on recent advances in malicious applications analysis and detection techniques for smartphones. 2012.

[47] M. Tavallaee, N. Stakhanova, and A. A. Ghorbani. Toward credible evaluation of anomaly-based intrusion-detection methods. *IEEE Transactions on Sys. Man Cyber Part C*, 40(5): 516–524, Sept. 2010.

[48] J. W. Ulvila and J. E. Gaffney. Evaluation of intrusion detection systems. *Journal of Research of the National Institute of Standards and Technology*, 108(6): 453–471, 2003.

[49] Q. Yan, Y. Li, T. Li, and R. Deng. Insights into malware detection and prevention on mobile phones. In *International Conference on Security Technology*, volume 58 of *Communications in Computer and Information Science*, pages 242–249. Springer Berlin Heidelberg, 2009.

[50] C. Zheng, S. Zhu, S. Dai, G. Gu, X. Gong, X. Han, and W. Zou. Smartdroid: An automatic system for revealing ui-based trigger conditions in android applications. In *Proceedings of the Second ACM Workshop on Security and Privacy in Smartphones and Mobile Devices*, SPSM '12, pages 93–104, New York, NY, USA, 2012. ACM.

[51] M. Zheng, M. Sun, and J. Lui. Droid analytics: A signature based analytic system to collect, extract, analyze and associate android malware. In *Trust, Security and Privacy in Computing and Communications (TrustCom), 2013 12th IEEE International Conference on*, pages 163–171. IEEE, 2013.

[52] Y. Zhou and X. Jiang. Dissecting Android malware: Characterization and evolution. In *IEEE Symposium on Security and Privacy (SP)*, pages 95–109. IEEE, 2012.

Appendix

List of the Reviewed Papers

Android Platform:	
Malware Detection	[1, 2, 3, 4, 5, 7, 8, 9, 10, 11, 12, 15, 16, 17, 20, 21, 22, 23, 25, 27, 28, 29, 30, 31, 34, 36, 38, 39, 40, 41, 42, 46, 48, 50, 51, 52, 53, 54, 56, 57, 58, 59, 60, 61, 62, 64, 65, 67, 68, 70, 71, 72, 75, 76, 77, 78, 80, 87, 93, 94, 95, 96, 98, 99, 100, 102, 104, 105, 110, 111, 116]
NIDS-HIDS	[6, 13, 43, 45, 47, 49, 69, 73, 85, 106]
NIDS	[35, 44, 66, 9]
HIDS	[14, 26, 32, 37, 63, 74, 79, 84, 86, 92, 101, 103]
Symbian OS:	
Malware Detection	[82, 107, 109, 112, 115]
HIDS	[108, 117, 119]
Windows Mobile OS:	
Malware Detection	[83, 89]
NIDS-HIDS	[118, 120]
HIDS	[88, 113]
Other Platforms:	
Malware Detection	[18, 90]
NIDS-HIDS	[24, 33, 55]
NIDS	[19, 81]
HIDS	[91, 114]

References

[1] Sanae Rosen, Zhiyun Qian, and Z Morely Mao. Appprofiler: a flexible method of exposing privacy-related behavior in android applications to end users. In *Proceedings of the third ACM conference on Data and application security and privacy*, pages 221–232. ACM, 2013.

[2] Saurabh Chakradeo, Bradley Reaves, Patrick Traynor, and William Enck. Mast: triage for market-scale mobile malware analysis. In *Proceedings of the sixth ACM conference on Security and privacy in wireless and mobile networks*, pages 13–24. ACM, 2013.

[3] Johannes Hoffmann, Martin Ussath, Thorsten Holz, and Michael Spreitzenbarth. Slicing droids: program slicing for smali code. In *Proceedings of the 28th Annual ACM Symposium on Applied Computing*, pages 1844–1851. ACM, 2013.

[4] Kevin Joshua Abela, Jan Raynier Delas Alas, Don Kristopher Angeles, Robert Joseph Tolentino, and Miguel Alberto Gomez. Automated malware detection for android amda. In *The Second International*

Conference on Cyber Security, Cyber Peacefare and Digital Forensic (CyberSec2013), pages 180–188. The Society of Digital Information and Wireless Communication, 2013.

[5] Zarni Aung and Win Zaw. Permission-based android malware detection. *International Journal of Scientific and Technology Research*, 2(3): 228–234, 2013.

[6] Michael Spreitzenbarth, Felix Freiling, Florian Echtler, Thomas Schreck, and Johannes Hoffmann. Mobile-sandbox: having a deeper look into android applications. In *Proceedings of the 28th Annual ACM Symposium on Applied Computing*, pages 1808–1815. ACM, 2013.

[7] Mo Ghorbanzadeh, Yang Chen, Zhongmin Ma, T Charles Clancy, and Robert McGwier. A neural network approach to category validation of android applications. In *Computing, Networking and Communications (ICNC), 2013 International Conference on*, pages 740–744. IEEE, 2013.

[8] Min Zheng, Mingshen Sun, and John Lui. Droid analytics: A signature based analytic system to collect, extract, analyze and associate android malware. In *Trust, Security and Privacy in Computing and Communications (TrustCom), 2013 12th IEEE International Conference on*, pages 163–171. IEEE, 2013.

[9] Borja Sanz, Igor Santos, Javier Nieves, Carlos Laorden, Inigo Alonso-Gonzalez, and Pablo G Bringas. Mads: Malicious android applications detection through string analysis. In *Network and System Security*, pages 178–191. Springer, 2013.

[10] Yibing Zhongyang, Zhi Xin, Bing Mao, and Li Xie. Droidalarm: an all-sided static analysis tool for android privilege-escalation malware. In *Proceedings of the 8th ACM SIGSAC symposium on Information, computer and communications security*, pages 353–358. ACM, 2013.

[11] Wu Zhou, Yajin Zhou, Michael Grace, Xuxian Jiang, and Shihong Zou. Fast, scalable detection of piggybacked mobile applications. In *Proceedings of the third ACM conference on Data and application security and privacy*, pages 185–196. ACM, 2013.

[12] Vaibhav Rastogi, Yan Chen, and William Enck. Appsplayground: Automatic security analysis of smartphone applications. In *Proceedings of the third ACM conference on Data and application security and privacy*, pages 209–220. ACM, 2013.

[13] Hiroki Kuzuno and Satoshi Tonami. Signature generation for sensitive information leakage in android applications. In *Data Engineering*

Workshops (ICDEW), 2013 IEEE 29th International Conference on, pages 112–119. IEEE, 2013.

[14] Johannes Hoffmann, Stephan Neumann, and Thorsten Holz. Mobile malware detection based on energy fingerprints a dead end? In *Research in Attacks, Intrusions, and Defenses*, pages 348–368. Springer, 2013.

[15] Yousra Aafer, Wenliang Du, and Heng Yin. Droidapiminer: Mining api-level features for robust malware detection in android. In *Security and Privacy in Communication Networks*, pages 86–103. Springer, 2013.

[16] John Demme, Matthew Maycock, Jared Schmitz, Adrian Tang, Adam Waksman, Simha Sethumadhavan, and Salvatore Stolfo. On the feasibility of online malware detection with performance counters. In *Proceedings of the 40th Annual International Symposium on Computer Architecture*, pages 559–570. ACM, 2013.

[17] Hugo Gascon, Fabian Yamaguchi, Daniel Arp, and Konrad Rieck. Structural detection of android malware using embedded call graphs. In *Proceedings of the 2013 ACM workshop on Artificial intelligence and security*, pages 45–54. ACM, 2013.

[18] Yuan Zhang, Min Yang, Bingquan Xu, Zhemin Yang, Guofei Gu, Peng Ning, X Sean Wang, and Binyu Zang. Vetting undesirable behaviors in android apps with permission use analysis. In *Proceedings of the 2013 ACM SIGSAC conference on Computer & communications security*, pages 611–622. ACM, 2013.

[19] Ruofan Jin and Bing Wang. Malware detection for mobile devices using software-defined networking. In *Research and Educational Experiment Workshop (GREE), 2013 Second GENI*, pages 81–88. IEEE, 2013.

[20] Hyo-Sik Ham and Mi-Jung Choi. Analysis of android malware detection performance using machine learning classifiers. In *ICT Convergence (ICTC), 2013 International Conference on*, pages 490–495. IEEE, 2013.

[21] Federico Maggi, Andrea Valdi, and Stefano Zanero. Andrototal: a flexible, scalable toolbox and service for testing mobile malware detectors. In *Proceedings of the Third ACM workshop on Security and privacy in smartphones & mobile devices*, pages 49–54. ACM, 2013.

[22] Suleiman Y Yerima, Sakir Sezer, Gavin McWilliams, and Igor Muttik. A new android malware detection approach using bayesian classification. In *Advanced Information Networking and Applications (AINA), 2013 IEEE 27th International Conference on*, pages 121–128. IEEE, 2013.

[23] Brandon Amos, Hamilton Turner, and Jules White. Applying machine learning classifiers to dynamic android malware detection at scale. In *Wireless Communications and Mobile Computing Conference (IWCMC), 2013 9th International*, pages 1666–1671. IEEE, 2013.

[24] Byungha Choi, Sung-Kyo Choi, and Kyungsan Cho. Detection of mobile botnet using vpn. In *Innovative Mobile and Internet Services in Ubiquitous Computing (IMIS), 2013 Seventh International Conference on*, pages 142–148. IEEE, 2013.

[25] Aiman A Abu Samra and Osama A Ghanem. Analysis of clustering technique in android malware detection. In *Innovative Mobile and Internet Services in Ubiquitous Computing (IMIS), 2013 Seventh International Conference on*, pages 729–733. IEEE, 2013.

[26] Bryan Dixon and Shivakant Mishra. Power based malicious code detection techniques for smartphones. In *Trust, Security and Privacy in Computing and Communications (TrustCom), 2013 12th IEEE International Conference on*, pages 142–149. IEEE, 2013.

[27] Parvez Faruki, Vijay Ganmoor, Vijay Laxmi, MS Gaur, and Ammar Bharmal. Androsimilar: robust statistical feature signature for android malware detection. In *Proceedings of the 6th International Conference on Security of Information and Networks*, pages 152–159. ACM, 2013.

[28] Kevin Joshua Abela, Don Kristopher Angeles, Jan Raynier Delas Alas, Robert Joseph Tolentino, and Miguel Alberto Gomez. An automated malware detection system for android using behavior-based analysis amda. *International Journal of Cyber-Security and Digital Forensics (IJCSDF)*, 2(2):1–11, 2013.

[29] Borja Sanz, Igor Santos, Xabier Ugarte-Pedrero, Carlos Laorden, Javier Nieves, and Pablo G Bringas. Instance-based anomaly method for android malware detection. pages 387–394, 2013.

[30] Anand Paturi, Manoj Cherukuri, John Donahue, and Srinivas Mukkamala. Mobile malware visual analytics and similarities of attack toolkits (malware gene analysis). In *Collaboration Technologies and Systems (CTS), 2013 International Conference on*, pages 149–154. IEEE, 2013.

[31] Heqing Huang, Sencun Zhu, Peng Liu, and Dinghao Wu. A framework for evaluating mobile app repackaging detection algorithms. In *Trust and Trustworthy Computing*, pages 169–186. Springer, 2013.

[32] Monica Curti, Alessio Merlo, Mauro Migliardi, and Simone Schiappacasse. Towards energy-aware intrusion detection systems on mobile

devices. In *High Performance Computing and Simulation (HPCS), 2013 International Conference on*, pages 289–296. IEEE, 2013.

[33] Wei Yu, Zhijiang Chen, Guobin Xu, Sixiao Wei, and Nnanna Ekedebe. A threat monitoring system for smart mobiles in enterprise networks. In *Proceedings of the 2013 Research in Adaptive and Convergent Systems*, pages 300–305. ACM, 2013.

[34] Jianlin Xu, Yifan Yu, Zhen Chen, Bin Cao, Wenyu Dong, Yu Guo, and Junwei Cao. Mobsafe: cloud computing based forensic analysis for massive mobile applications using data mining. *Tsinghua Science and Technology*, 18(4), 2013.

[35] Lena Tenenboim-Chekina, Lior Rokach, and Bracha Shapira. Ensemble of feature chains for anomaly detection. In *Multiple Classifier Systems*, pages 295–306. Springer, 2013.

[36] Yibing Zhongyang, Zhi Xin, Bing Mao, and Li Xie. Droidalarm: an all-sided static analysis tool for android privilege-escalation malware. In *Proceedings of the 8th ACM SIGSAC symposium on Information, computer and communications security*, pages 353–358. ACM, 2013.

[37] Hwan-Taek Lee, Minkyu Park, and Seong-Je Cho. Detection and pre vention of lena malware on android. *Journal of Internet Services and Information Security (JISIS)*, 3(3/4):63–71, 2013.

[38] Seung-Hyun Seo, Aditi Gupta, Asmaa Mohamed Sallam, Elisa Bertino, and Kangbin Yim. Detecting mobile malware threats to home-land security through static analysis. *Journal of Network and Computer Applications*, 38:43–53, 2013.

[39] Byeongho Kang, BooJoong Kang, Jungtae Kim, and Eul Gyu Im. Android malware classification method: Dalvik bytecode frequency analysis. In *Proceedings of the 2013 Research in Adaptive and Convergent Systems*, pages 349–350. ACM, 2013.

[40] Suyeon Lee, Jehyun Lee, and Heejo Lee. Screening smartphone appli-cations using behavioral signatures. In *Security and Privacy Protection in Information Processing Systems*, pages 14–27. Springer, 2013.

[41] Parvez Faruki, Vijay Laxmi, Vijay Ganmoor, MS Gaur, and Ammar Bharmal. Droidolytics: Robust feature signature for repackaged android apps on official and third party android markets. In *Advanced Com-puting, Networking and Security (ADCONS), 2013 2nd International Conference on*, pages 247–252. IEEE, 2013.

[42] Shaoyin Cheng, Shengmei Luo, Zifeng Li, Wei Wang, Yan Wu, and Fan Jiang. Static detection of dangerous behaviors in android apps. In *Cyberspace Safety and Security*, pages 363–376. Springer, 2013.

[43] Zhizhong Wu, Xuehai Zhou, and Jun Xu. A result fusion based distributed anomaly detection system for android smartphones. *Journal of Networks*, 8(2), 2013.

[44] Ryan Johnson, Zhaohui Wang, Angelos Stavrou, and Jeff Voas. Exposing software security and availability risks for commercial mobile devices. In *Reliability and Maintainability Symposium (RAMS), 2013 Proceedings-Annual*, pages 1–7. IEEE, 2013.

[45] K Saritha and R Samaiah. Behavior analysis of mobile system in cloud computing. In *International Journal of Engineering Research and Technology*, volume 2. ESRSA Publications, 2013.

[46] Thomas Eder, Michael Rodler, Dieter Vymazal, and Markus Zeilinger. Ananas-a framework for analyzing android applications. In *Availability, Reliability and Security (ARES), 2013 Eighth International Conference on*, pages 711–719. IEEE, 2013.

[47] Roshanak Roshandel, Payman Arabshahi, and Radha Poovendran. Lidar: a layered intrusion detection and remediationframework for smartphones. In *Proceedings of the 4th international ACM Sigsoft symposium on Architecting critical systems*, pages 27–32. ACM, 2013.

[48] Mohammad Karami, Mohamed Elsabagh, Parnian Najafiborazjani, and Angelos Stavrou. Behavioral analysis of android applications using automated instrumentation. In *Software Security and Reliability-Companion (SERE-C), 2013 IEEE 7th International Conference on*, pages 182–187. IEEE, 2013.

[49] Fangfang Yuan, Lidong Zhai, Yanan Cao, and Li Guo. Research of intrusion detection system on android. In *Services (SERVICES), 203 IEEE Ninth World Congress on*, pages 312–316. IEEE, 2013.

[50] Ryo Sato, Daiki Chiba, and Shigeki Goto. Detecting android malware by analyzing manifest files. *Proceedings of the Asia-Pacific Advanced Network*, 36: 23–31, 2013.

[51] Dong-uk Kim, Jeongtae Kim, and Sehun Kim. A malicious application detection framework using automatic feature extraction tool on android market. In *3rd International Conference on Computer Science and Information Technology (ICCSIT'2013)*, pages 4–5, 2013.

[52] Jonathan Crussell, Clint Gibler, and Hao Chen. *Scalable semantics-based detection of similar Android applications*. ESORICS, 2013.

[53] Veelasha Moonsamy, Jia Rong, and Shaowu Liu. Mining permission patterns for contrasting clean and malicious android applications. *Future Generation Computer Systems*, 2013.

[54] You Joung Ham, Hyung-Woo Lee, Jae Deok Lim, and Jeong Nyeo Kim. Droidvulmonandroid based mobile device vulnerability analysis and monitoring system. In *Next Generation Mobile Apps, Services and Technologies (NGMAST), 2013 Seventh International Conference on*, pages 26–31. IEEE, 2013.

[55] Ilona Murynets and Roger Piqueras Jover. Anomaly detection in cellular machine-to-machine communications. In *Communications (ICC), 2013 IEEE International Conference on*, pages 2138–2143. IEEE, 2013.

[56] Zhemin Yang, Min Yang, Yuan Zhang, Guofei Gu, Peng Ning, and X Sean Wang. Appintent: Analyzing sensitive data transmission in android for privacy leakage detection. In *Proceedings of the 2013 ACM SIGSAC conference on Computer & communications security*, pages 1043–1054. ACM, 2013.

[57] Wei Xu, Fangfang Zhang, and Sencun Zhu. Permlyzer: Analyzing permission usage in android applications. In *Software Reliability Engineering (ISSRE), 2013 IEEE 24th International Symposium on*, pages 400–410. IEEE, 2013.

[58] Steve Hanna, Ling Huang, Edward Wu, Saung Li, Charles Chen, and Dawn Song. Juxtapp: A scalable system for detecting code reuse among android applications. In *Detection of Intrusions and Malware, and Vulnerability Assessment*, pages 62–81. Springer, 2013.

[59] Dong-Jie Wu, Ching-Hao Mao, Te-En Wei, Hahn-Ming Lee, and Kuo-Ping Wu. Droidmat: Android malware detection through manifest and api calls tracing. In *Information Security (Asia JCIS), 2012 Seventh Asia Joint Conference on*, pages 62–69. IEEE, 2012.

[60] Hao Peng, Chris Gates, Bhaskar Sarma, Ninghui Li, Yuan Qi, Rahul Potharaju, Cristina Nita-Rotaru, and Ian Molloy. Using probabilistic generative models for ranking risks of android apps. In *Proceedings of the 2012 ACM conference on Computer and communications security*, pages 241–252. ACM, 2012.

[61] Yajin Zhou, Zhi Wang, Wu Zhou, and Xuxian Jiang. Hey, you, get off of my market: Detecting malicious apps in official and alternative android markets. In *Proceedings of the 19th Annual Network and Distributed System Security Symposium*, pages 5–8, 2012.

[62] Wu Zhou, Yajin Zhou, Xuxian Jiang, and Peng Ning. Detecting repackaged smartphone applications in third-party android marketplaces. In *Proceedings of the second ACM conference on Data and Application Security and Privacy*, pages 317–326. ACM, 2012.

[63] Asaf Shabtai, Uri Kanonov, Yuval Elovici, Chanan Glezer, and Yael Weiss. "andromaly": a behavioral malware detection framework for android devices. *Journal of Intelligent Information Systems*, 38(1): 161–190, 2012.

[64] Axelle Apvrille and Tim Strazzere. Reducing the window of opportunity for android malware gotta catch'em all. *Journal in Computer Virology*, 8(1–2):61–71, 2012.

[65] Abhijith Shastry, Murat Kantarcioglu, Yan Zhou, and Bhavani Thuraisingham. Randomizing smartphone malware profiles against statistical mining techniques. In *Data and Applications Security and Privacy XXVI*, pages 239–254. Springer, 2012.

[66] Te-En Wei, Ching-Hao Mao, Albert B Jeng, Hahn-Ming Lee, Horng-Tzer Wang, and Dong-Jie Wu. Android malware detection via a latent network behavior analysis. In *Trust, Security and Privacy in Computing and Communications (TrustCom), 2012 IEEE 11th International Conference on*, pages 1251–1258. IEEE, 2012.

[67] Borja Sanz, Igor Santos, Carlos Laorden, Xabier Ugarte-Pedrero, PabloGarcia Bringas, and Gonzalo Álvarez. Puma: Permission usage to detect malware in android. In Álvaro Herrero, Václav Snášel, Ajith Abraham, Ivan Zelinka, Bruno Baruque, Héctor Quintián, José Luis Calvo, Javier Sedano, and Emilio Corchado, editors, *International Joint Conference CISIS12-ICEUTE12-SOCO´12 Special Sessions*, volume 189 of *Advances in Intelligent Systems and Computing*, pages 289–298. Springer Berlin Heidelberg, 2013.

[68] Chao Yang, Vinod Yegneswaran, Phillip Porras, and Guofei Gu. Detecting money-stealing apps in alternative android markets. In *Proceedings of the 2012 ACM conference on Computer and communications security*, pages 1034–1036. ACM, 2012.

[69] Ingo Bente, Bastian Hellmann, Joerg Vieweg, Josef von Helden, and Gabi Dreo. Tcads: Trustworthy, context-related anomaly detection for smartphones. In *Network-Based Information Systems (NBiS), 2012 15th International Conference on*, pages 247–254. IEEE, 2012.

[70] Justin Sahs and Latifur Khan. A machine learning approach to android malware detection. In *Intelligence and Security Informatics Conference (EISIC), 2012 European*, pages 141–147. IEEE, 2012.

[71] Seung-Hyun Seo, Dong-Guen Lee, and Kangbin Yim. Analysis on maliciousness for mobile applications. In *Innovative Mobile and Internet Services in Ubiquitous Computing (IMIS), 2012 Sixth International Conference on*, pages 126–129. IEEE, 2012.

[72] Lena Chekina, Duku Mimran, Lior Rokach, Yuval Elovici, and Bracha Shapira. Detection of deviations in mobile applications network behavior. *arXiv preprint arXiv:1208.0564*, 2012.

[73] Muhamed Halilovic and Abdulhamit Subasi. Intrusion detection on smartphones. *arXiv preprint arXiv:1211.6610*, 2012.

[74] PENG Guojun, SHAO Yuru, WANG Taige, ZHAN Xian, and ZHANG Huanguo. Research on android malware detection and interception based on behavior monitoring. 17(5), 2012.

[75] Cong Zheng, Shixiong Zhu, Shuaifu Dai, Guofei Gu, Xiaorui Gong, Xinhui Han, and Wei Zou. Smartdroid: an automatic system for revealing ui-based trigger conditions in android applications. In *Proceedings of the second ACM workshop on Security and privacy in smartphones and mobile devices*, pages 93–104. ACM, 2012.

[76] Gianluca Dini, Fabio Martinelli, Ilaria Matteucci, Marinella Petrocchi, Andrea Saracino, and Daniele Sgandurra. A multi-criteria-based evaluation of android applications. In *Trusted Systems*, pages 67–82. Springer, 2012.

[77] Michael Grace, Yajin Zhou, Zhi Wang, and Xuxian Jiang. Systematic detection of capability leaks in stock android smartphones. In *Proceedings of the 19th Annual Symposium on Network and Distributed System Security*, 2012.

[78] Michael Grace, Yajin Zhou, Qiang Zhang, Shihong Zou, and Xuxian Jiang. Riskranker: scalable and accurate zero-day android malware detection. In *Proceedings of the 10th international conference on Mobile systems, applications, and services*, pages 281–294. ACM, 2012.

[79] Gianluca Dini, Fabio Martinelli, Andrea Saracino, and Daniele Sgandurra. Madam: a multi-level anomaly detector for android malware. In *Computer Network Security*, pages 240–253. Springer, 2012.

[80] Lingguang Lei, Yuewu Wang, Jiwu Jing, Zhongwen Zhang, and Xingjie Yu. Meaddroid: detecting monetary theft attacks in android by dvm monitoring. In *Information Security and Cryptology–ICISC 2012*, pages 78–91. Springer, 2013.

[81] Jordi Cucurull, Simin Nadjm-Tehrani, and Massimiliano Raciti. Modular anomaly detection for smartphone ad hoc communication. In *Information Security Technology for Applications*, pages 65–81. Springer, 2012.

[82] Kejun Xin, Gang Li, Zhongyuan Qin, and Qunfang Zhang. Malware detection in smartphone using hidden markov model. In *Multimedia Information Networking and Security (MINES), 2012 Fourth International Conference on*, pages 857–860. IEEE, 2012.

[83] Hua Zha and Chunlin Peng. Method of smartphone users' information protection based on composite behavior monitor. In *Intelligent Computing Technology*, pages 252–259. Springer, 2012.

[84] Chanmin Yoon, Dongwon Kim, Wonwoo Jung, Chulkoo Kang, and Hojung Cha. Appscope: Application energy metering framework for android smartphone using kernel activity monitoring. In *USENIX ATC*, 2012.

[85] You-Joung Ham, Won-Bin Choi, Hyung-Woo Lee, JaeDeok Lim, and Jeong Nyeo Kim. Vulnerability monitoring mechanism in android based smartphone with correlation analysis on event-driven activities. In *Computer Science and Network Technology (ICCSNT), 2012 2nd International Conference on*, pages 371–375. IEEE, 2012.

[86] Lok Kwong Yan and Heng Yin. Droidscope: seamlessly reconstructing the os and dalvik semantic views for dynamic android malware analysis. In *Proceedings of the 21st USENIX Security Symposium*, 2012.

[87] Iker Burguera, Urko Zurutuza, and Simin Nadjm-Tehrani. Crowdroid: behavior-based malware detection system for android. In *Proceedings of the 1st ACM workshop on Security and privacy in smartphones and mobile devices*, pages 15–26. ACM, 2011.

[88] Hahnsang Kim, Kang G Shin, and Padmanabhan Pillai. Modelz: monitoring, detection, and analysis of energy-greedy anomalies in mobile handsets. *Mobile Computing, IEEE Transactions on*, 10(7): 968–981, 2011.

[89] Hsiu-Sen Chiang and W Tsaur. Identifying smartphone malware using data mining technology. In *Computer Communications and Networks (ICCCN), 2011 Proceedings of 20th International Conference on*, pages 1–6. IEEE, 2011.

[90] Bryan Dixon, Yifei Jiang, Abhishek Jaiantilal, and Shivakant Mishra. Location based power analysis to detect malicious code in smartphones. In *Proceedings of the 1st ACM workshop on Security and privacy in smartphones and mobile devices*, pages 27–32. ACM, 2011.

[91] Zhang Lei, Zhu Junmao, Tian Zhongguang, Liu Yulong, and Wang Tao. Design of mobile phone security system based on detection of abnormal

behavior. In *Proceedings of the 2011 First International Conference on Instrumentation, Measurement, Computer, Communication and Control*, pages 479–482. IEEE Computer Society, 2011.

[92] Amir Houmansadr, Saman A Zonouz, and Robin Berthier. A cloud-based intrusion detection and response system for mobile phones. In *Dependable Systems and Networks Workshops (DSN-W), 2011 IEEE/IFIP 41st International Conference on*, pages 31–32. IEEE, 2011.

[93] Peter Gilbert, Byung-Gon Chun, Landon P Cox, and Jaeyeon Jung. Vision: automated security validation of mobile apps at app markets. In *Proceedings of the second international workshop on Mobile cloud computing and services*, pages 21–26. ACM, 2011.

[94] Erika Chin, Adrienne Porter Felt, Kate Greenwood, and David Wagner. Analyzing inter-application communication in android. In *Proceedings of the 9th international conference on Mobile systems, applications, and services*, pages 239–252. ACM, 2011.

[95] Leonid Batyuk, Markus Herpich, Seyit Ahmet Camtepe, Karsten Raddatz, A-D Schmidt, and Sahin Albayrak. Using static analysis for automatic assessment and mitigation of unwanted and malicious activities within android applications. In *Malicious and Unwanted Software (MALWARE), 2011 6th International Conference on*, pages 66–72. IEEE, 2011.

[96] Takamasa Isohara, Keisuke Takemori, and Ayumu Kubota. Kernel-based behavior analysis for android malware detection. In *Computational Intelligence and Security (CIS), 2011 Seventh International Conference on*, pages 1011–1015. IEEE, 2011.

[97] Lei Liu and Dai Ping Li. Analysis based on of android malicious code intrusion detection. *Advanced Materials Research*, 756: 3924–3928, 2013.

[98] Francesco Di Cerbo, Andrea Girardello, Florian Michahelles, and Svetlana Voronkova. Detection of malicious applications on android os. In *Computational Forensics*, pages 138–149. Springer, 2011.

[99] Liang Xie, Xinwen Zhang, Jean-Pierre Seifert, and Sencun Zhu. pbmds: a behavior-based malware detection system for cellphone devices. In *Proceedings of the third ACM conference on Wireless network security*, pages 37–48. ACM, 2010.

[100] Markus Jakobsson and Karl-Anders Johansson. Retroactive detection of malware with applications to mobile platforms. In *Proceedings of the 5th USENIX conference on Hot topics in security*, pages 1–13. USENIX Association, 2010.

[101] Asaf Shabtai, Uri Kanonov, and Yuval Elovici. Intrusion detection for mobile devices using the knowledge-based, temporal abstraction method. *Journal of Systems and Software*, 83(8): 1524–1537, 2010.

[102] Thomas Blasing, Leonid Batyuk, A-D Schmidt, Seyit Ahmet Camtepe, and Sahin Albayrak.   An android application sandbox system for suspicious software detection. In *Malicious and Unwanted Software (MALWARE), 2010 5th International Conference on*, pages 55–62. IEEE, 2010.

[103] Asaf Shabtai and Yuval Elovici.   Applying behavioral detection on android-based devices.   In *Mobile Wireless Middleware, Operating Systems, and Applications*, pages 235–249. Springer, 2010.

[104] Asaf Shabtai, Yuval Fledel, and Yuval Elovici. Automated static code analysis for classifying android applications using machine learning. In *Computational Intelligence and Security (CIS), 2010 International Conference on*, pages 329–333. IEEE, 2010.

[105] William Enck, Peter Gilbert, Byung-Gon Chun, Landon P Cox, Jaeyeon Jung, Patrick McDaniel, and Anmol Sheth. Taintdroid: An information-flow tracking system for realtime privacy monitoring on smartphones. In *OSDI*, volume 10, pages 1–6, 2010.

[106] Georgios Portokalidis, Philip Homburg, Kostas Anagnostakis, and Herbert Bos. Paranoid android: versatile protection for smartphones. In *Proceedings of the 26th Annual Computer Security Applications Conference*, pages 347–356. ACM, 2010.

[107] Tansu Alpcan, Christian Bauckhage, and Aubrey-Derrick Schmidt. A probabilistic diffusion scheme for anomaly detection on smartphones. In *Information Security Theory and Practices. Security and Privacy of Pervasive Systems and Smart Devices*, pages 31–46. Springer, 2010.

[108] Fudong Li, Nathan Clarke, Maria Papadaki, and Paul Dowland. Behaviour profiling on mobile devices. In *Emerging Security Technologies (EST), 2010 International Conference on*, pages 77–82. IEEE, 2010.

[109] Ashkan Sharifi Shamili, Christian Bauckhage, and Tansu Alpcan. Malware detection on mobile devices using distributed machine learning.   In *Pattern Recognition (ICPR), 2010 20th International Conference on*, pages 4348–4351. IEEE, 2010.

[110] William Enck, Machigar Ongtang, and Patrick McDaniel. On lightweight mobile phone application certification. In *Proceedings of the 16th ACM conference on Computer and communications security*, pages 235–245. ACM, 2009.

[111] A-D Schmidt, Rainer Bye, H-G Schmidt, Jan Clausen, Osman Kiraz, Kamer A Yuksel, Seyit Ahmet Camtepe, and Sahin Albayrak. Static analysis of executables for collaborative malware detection on android. In *Communications, 2009. ICC'09. IEEE International Conference on*, pages 1–5. IEEE, 2009.

[112] Lei Liu, Guanhua Yan, Xinwen Zhang, and Songqing Chen. Virusmeter: Preventing your cellphone from spies. In *Recent Advances in Intrusion Detection*, pages 244–264. Springer, 2009.

[113] Jong-seok Lee, Tae-Hyung Kim, and Jong Kim. Energy-efficient run-time detection of malware-infected executables and dynamic libraries on mobile devices. In *Future Dependable Distributed Systems, 2009 Software Technologies for*, pages 143–149. IEEE, 2009.

[114] Aubrey-Derrick Schmidt, Frank Peters, Florian Lamour, Christian Scheel, Seyit Ahmet Çamtepe, and Şahin Albayrak. Monitoring smartphones for anomaly detection. *Mobile Networks and Applications*, 14(1): 92–106, 2009.

[115] Liang Xie, Xinwen Zhang, Ashwin Chaugule, Trent Jaeger, and Sencun Zhu. Designing system-level defenses against cellphone malware. In *Reliable Distributed Systems, 2009. SRDS'09. 28th IEEE International Symposium on*, pages 83–90. IEEE, 2009.

[116] Aubrey-Derrick Schmidt, Hans-Gunther Schmidt, Jan Clausen, Kamer A Yuksel, Osman Kiraz, Ahmet Camtepe, and Sahin Albayrak. Enhancing security of linux-based android devices. In *in Proceedings of 15th International Linux Kongress. Lehmann*, 2008.

[117] Abhijit Bose, Xin Hu, Kang G Shin, and Taejoon Park. Behavioral detection of malware on mobile handsets. In *Proceedings of the 6th international conference on Mobile systems, applications, and services*, pages 225–238. ACM, 2008.

[118] Hahnsang Kim, Joshua Smith, and Kang G Shin. Detecting energy-greedy anomalies and mobile malware variants. In *Proceedings of the 6th international conference on Mobile systems, applications, and services*, pages 239–252. ACM, 2008.

[119] Deepak Venugopal and Guoning Hu. Efficient signature based malware detection on mobile devices. *Mobile Information Systems*, 4(1): 33–49, 2008.

[120] Timothy K Buennemeyer, Theresa M Nelson, Lee M Clagett, John Paul Dunning, Randy C Marchany, and Joseph G Tront. Mobile device profiling and intrusion detection using smart batteries. In *Hawaii International Conference on System Sciences, Proceedings of the 41st Annual*, pages 296–296. IEEE, 2008.

Biographies

Abdullah Alzahrani is a PhD candidate at the faculty of Computer Science, University of New Brunswick, Canada. He is a lecturer at Computer Science and computer Engineering department, University of Hail, Saudi Arabia. His research interests include botnet detection, Android security, network security, malware analysis and reverse engineering. He is a member of the Information Security Centre of Excellence, University of New Brunswick.

Dr. Natalia Stakhanova is the New Brunswick Innovation Research Chair in Cyber Security at University of New Brunswick, Canada. Her research interests include intrusion detection and response, smartphone security, security assessment and generally network and computer security. Natalia Stakhanova was the recipient of the Nokia Best Student Paper Award at The IEEE

International Conference on Advanced Information Networking and Applications (AINA). She served on the program committee of several conferences and workshops in area of information security and assurance, including the Conference on Privacy, Security and Trust (PST). Natalia developed a number of technologies that have been adopted by high-tech companies such as IBM and she currently has three patents in the field of computer security.

Hugo Gonzalez is a PhD student at the Information Security Centre of Excellence, University of New Brunswick, Canada. He is a faculty member of the Polytechnic University of San Luis Potosi, Mexico. His current research interests include network security and malware analysis. He is a member of the Association for Computing Machinery, the IEEE Computer Society and The Honeynet Project.

Ali Ghorbani has held a variety of positions in academia for the past 34 years. He currently serves as Dean of the Faculty of Computer Science and Founding Director of the Information Security Centre of Excellence at the

University of New Brunswick (UNB), Fredericton, Canada. Dr. Ghorbani is the co-Editor-In-Chief of Computational Intelligence, an international journal. He supervised more than 150 research associates, postdoctoral fellows, and undergraduate & graduate students and authored more than 250 research papers in journals and conference proceedings and has edited 11 volumes. He is the co-inventor of 3 patents in the area of Network Security and Web Intelligence. In 2012 he spawn off "Ara Labs Security Solutions" and "Eyesover Technologies". His current research focus is Network & Information Security, Complex Adaptive Systems, Critical Infrastructure Protection, and Web Intelligence. His book, Intrusion Detection and Prevention Systems: Concepts and Techniques, published by Springer in October 2009. Dr. Ghorbani is the Senior member of IEEE and the member of ACM, and Canadian Information Processing Society (CIPS). He is also the coordinator of the Privacy, Security and Trust (PST*net) research network.

Enabling Wireless Sensor Nodes for Self-Contained Jamming Detection

Stephan Kornemann[1], Steffen Ortmann[1], Peter Langendörfer[1] and Alexandros Fragkiadakis[2]

[1] *IHP, Im Technologiepark 25, D-15236 Frankfurt (Oder), Germany*
[2] *Institute of Computer Science, Foundation for Research and Technology-Hellas(FORTH), Heraklion, Crete*

Received 1 March 2014; Accepted 15 April 2014;
Publication 2 July 2014

Abstract

Jamming is an easy to execute attack to which wireless sensor networks are extremely vulnerable. If the application requires reliability, jamming needs to be detected and reported in order to cope with this attack. In this article, we investigate different approaches to identify jamming. Available jamming detection schemes primarily suffer from the usage of fixed thresholds as well as required effort. We adapted a variance-based estimate of signal-to-noise ratio measurements, called significance analysis, to the minor resources and computing efforts of wireless sensor nodes. As a start, we used real measurement data for theoretical analysis of the methods under investigation. Independently of the location of the jamming device, our significance analysis approach provides an immediate indication of jamming and can in theory be run with almost least effort, i.e., with O(14). On top of that, we implemented this approach on our state of the art sensor node and tested it in a real world outdoor setting. Our jamming detection engine monitors the wireless channel with a sampling rate of 10 ms. It returns a jamming detection decision within less than 5 ms while though achieving a detection accuracy in between 84 to 99 percent.

Keywords: Jamming detection, Wireless sensor networks, Security.

Journal of Cyber Security, Vol. 3 No. 2, 133–158.
doi: 10.13052/jcsm2245-1439.322

1 Introduction

Wireless sensor networks (WSNs) are more and more considered as a basis for new applications e.g. in the area of automation control or critical infrastructure protection. Such applications require a significant level of reliability. Jamming is an attack which needs to be considered as extremely dangerous. It can be easily executed by anybody since it does not need any detailed knowledge about the system to be attacked [10] nor expensive equipment. In addition its effect is significant since it immediately distorts the expected system behaviour.

Many projects [2–5, 11, 15–16] have proven fixed thresholds to be unsuitable for jamming detection in wireless networks no matter which channel characteristic is monitored. Beside physical conditions around the node, the distance to the jammer predominantly influences the changes in channel characteristics by jamming, e.g. the signal strength of the jammer obviously decreases with the distance to network under attack but still remains noticeable. Such behaviour can also be seen in Figures 2(a) and 2(b) in the next chapter. Since the location of a jammer is hardly to predict [7], sensor nodes cannot be pre-configured for reliable jamming detection, even if the jamming characteristic is known. Instead, sensor nodes must learn distinguishing regular from irregular (jamming) channel conditions. This is a difficult task since normal operation within the WSN, e.g., during contention phases, can look like jamming and by that cause false positives.

The contributions of this papers are:

- Introduction of different mathematical approaches which are suitable to detect jamming without prior knowledge on thresholds, i.e., self adaptive approaches.
- Analysis of the computational complexity of these approaches in order to evaluate whether or not they can be used on resource constraint wireless sensor nodes.
- Theoretical evaluation of a suitable metric, called significance analysis, using real measurement data.
- Realization and validation of significance analysis for real-time detection of jamming on wireless sensor nodes in practice.
- Discussion of approaches to reduce the number of false positives and false negatives.

In the next section we overview existing approaches and assess application of those in WSNs. Section 3 analyses and compares computational and memory effort of feasible solutions. Based on that, we introduce our low effort approach for online jamming indication. Section 4 evaluates simulation results based of data taken from real measurements in the laboratory. Based on that, we present implementation detail as well as achieved detection rates of our significance analysis on state of the art sensor nodes. Finally, we discuss trouble-shooting issues and provide concluding remarks.

2 Related Work

Applying wireless communication, WSNs by nature are vulnerable to disturbances or even blockages of the wireless channel. Such interference may of course occur due to changes in environmental physics, such as objects moving or changing weather conditions, especially humidity and rain [1]. However, interference on wireless channels can also be willingly provoked by an attacker. Such behaviour is usually called jamming. Many different jamming models as well as respective counter measures have been researched in depth in the past. Among others, random, reactive [16], periodic [4, 8] or constant [14] jamming models are applied for both single and multi-channel [9] attacks. Analysis of existing models in unison have agreed reactive, either random or periodic, jamming models to be the most effective ones for WSNs due to their jamming performance and energy efficiency.

Almost the same number of different counter measures using different basic metrics have been proposed. In the following we overview and assess the usage of these metrics. Xu et al. [16] evaluated the usage of different packet-based metrics. The Packet Send Ratio (PSR) basically indicates the number of packets sent during certain time period. This is equivalent to the analysis of time needed to access the wireless channel, usually given as Carrier Sensing Time (CST) when using CSMA MAC protocols. Thereby, a node detects a possible jamming attack if it can only access the channel with packet rates below certain threshold. This approach certainly performs poor if jamming and channel characteristics are unknown and hence, suitable thresholds cannot be fixed. Xu et al. [16] and Cakiroglu et al. [3] further have tested the packet-based metrics Packet Delivery Ratio (PDR) and Bad Packet Ratio (BPR). PDR counts the number of packets successfully sent or the number of acknowledgements received respectively. Jamming may then be detected if the ratio of successfully transmitted packets in percentage falls below certain threshold. Despite suitable processing effort this metrics implies usage of acknowledgement-based communication, which causes significant effort especially in case of good channel conditions. By that it cannot be expected suitable for resource constraint devices like WSNs. Similarly to the

PDR-based method, the BPR-based approach calculates the percental ratio between correct (good) and erroneous (bad) packets received. Obviously, this method can be used at the receiver side only. However, these approaches cannot reflect usage of WSNs in heavily interfered environments where packet losses result from ordinary operation. Both approaches further suffer from the usage of predefined thresholds. Similar problems and a substantial effort prevent from using Bit Error Rate (BER) as metric as proposed by Strasser et al. [12].

The metric best reflecting the physical conditions of the wireless channel is based on the Received Signal Strength (RSS). Using RSS the ratio between received signal power and received noise level, usually called Signal-to-Noise-Ratio (SNR), can be determined. From our point of view, SNR is the most suitable metric for jamming detection since any interference, whether or not caused by jamming, is reflected by changes of SNR. The challenge of analysing SNR for jamming detection is distinguishing abnormal (or anomalous) SNR from usual behaviour by assessing the actual SNR value. Jamming detection techniques based on SNR for wireless networks have been proven functional in several projects. Cabrera et. al. [2] determined a threshold-based jamming detection scheme for application in MANETs, called anomaly index. This approach probabilistically rates SNR drops at single nodes in a distributed decision process at respective cluster heads but still requires to set jamming thresholds at the cluster head. To get rid of the necessity to configure fixed thresholds, which is infeasible due to the unknown location of the jammer and unforeseeable channel characteristics, the network must "learn" usual SNR by itself. The most common SNR-based approaches exploit the standard deviation [11] or the variance of SNR readings [4, 13] to rate the occurrence of interference based on previous trend of SNR. Despite partially outstanding detection performance, these approaches cannot be used in sensor networks due to effort required, as we will show in the next section. We will also introduce our significance analysis approach that circumvents presented disadvantages when using variance of SNR as metric for jamming detection.

3 Math & Computational Effort

To determine jamming by changing SNR requires to contrast actual SNR values with expected ones. We indicate the significance of changes in the SNR as the degree of deviation from the expected range of values learnt from previous trend. This expected range of values is determined by the average of previous values to a lesser or greater degree of the variance of those. Determining the average and the variance of previous readings is not very complex according to mathematics, but it originally is unsuitable for sensor networks due to the calculation and memory effort. This especially holds true when it is used for jamming detection, where every new SNR value has to be processed immediately. Hence, the effort for processing one single SNR value is to be determined. In the following, we introduce the math and respective effort in detail and point out the main drawbacks. We finally show how to adapt these calculations to the needs of limited devices like sensor nodes.

3.1 Standard Variance

The variance indicates the range of values where the next reading is most likely in. We propose to apply a maximum likelihood estimate to determine the variance of previous readings σ_m, see Equation 1. The variance originally requires to use all previous (here n) readings for calculating the average value x and the differences of x to all n previous readings x_i. In summary, processing a new reading requires 2n additions, n subtractions and squares, plus one division and one root extraction operation. Hence, the calculation effort of a single run is $O(4n + 2)$, which finally equals to $O(n)$. In addition, the standard variance requires a large amount of memory due to the necessity to store all previous n values. Even though we do not consider the additional overhead required for memory access, the standard variance is unsuitable for sensor networks due to both calculation and memory efforts. Since we intend using the variance of the SNR for online jamming detection, both parameters need to be significantly reduced to be applicable for sensor networks. Hence, we adapted the standard calculation to sensor needs by applying the parallel axis theorem and a customised sliding window derivative.

$$\sigma_m = \sqrt{\sum_{i=1}^{n} (x_i - \bar{x})^2}; \quad \bar{x} = \frac{1}{n} \sum_{i=1}^{n} (x_i) \tag{1}$$

3.2 Reducing Calculation and Memory Effort

Obviously, on the fly processing of SNR readings must be very fast and may not depend on the number of values (n) used. The parallel axis theorem substitutes x by the average formula given in Equation 1. It allows processing consecutive sensor readings without the need to have all previous values available. Instead, only the sum of measurements and the sum of measurements squares need to be refreshed and stored, see Equation 2.

$$\sigma_m = \sqrt{(\frac{1}{n}\sum_{i=1}^{n} x_i^2) - \bar{x}^2} = \sqrt{(\frac{1}{n}\sum_{i=1}^{n} x_i^2) - (\frac{1}{n}\sum_{i=1}^{n} x_i)^2} \qquad (2)$$

Applying the parallel axis theorem fixes the calculation effort to eight operations $O(8)$ per processing. It only requires to 2 additions, 2 divisions and 2 squares plus one subtraction and one root extraction. Even the memory effort is very low. It requires to store three numbers only, i.e., the sum of measurements, the sum of measurements squares and the entire number of processed readings. Unfortunately, these sums are the main drawback of this approach. The SNR usually varies below 100 *dB* and hence, the squares of SNR readings may be very huge. Further the number of processed SNRs can rapidly increase and result in huge sums used for calculating the variance with the parallel axis theorem. The microcontroller used on sensor nodes usually apply a 16 bit or a 20 bit architecture. Due to its energy efficiency we use the 16 bit MSP430 microcontroller from Texas Instruments on our nodes. Consequently, the 16 bit architecture limits the size of the sums to 65536.

3.3 Adaptation to Sensor Node Needs

To keep the sums adequately low, we propose to include only a certain number of previous readings specified by an adaptable sliding window s in the calculation. The sliding window approach provides two benefits. It allows to influence the size of the sums as well as to properly adapt the number of considered measurements to the application. For example, SNRs of past days may be not of interest for jamming detection, whereas the readings of the last ten minutes may be much more important for comparison and evaluation. Note, we also have considered reducing the SNR value by a factor before processing, e.g., dividing SNR values by 10 or 100. By that, we may unnecessarily introduce processing overhead since we cannot guarantee the availability of hardware-accelerated floating point arithmetic.

Obviously, the sliding window approach falls back into $O(s)$ computational effort and requires to store s numbers in memory. Applying the parallel axis theorem allows to get rid of stored measurements as already shown, but is in principle unsuitable for the sliding window method, which requires these measurements. To allow efficient processing on sensor nodes, we propose combining both methods in σ_s by estimating the measurements within the

Table 1 Comparison of computational and memory effort of approaches introduced. Despite of best effort, rapidly growing sums prevent from using the parallel axis theorem on 16 or 20 bit architectures

Method	Calculation	Memory
Variance	O(4n+2) = O(n)	n numbers
Parallel Axis Theorem	O(8)	2 numbers
Significance indicator	O(14)	3 numbers

sliding window. The important values of the parallel axis theorem are the sum of measurements and the sum of measurement squares, as mentioned. Originally, these sums are updated at every sensing interval by processing on the next sensor reading. To apply the sliding window method, the current measurement is added to the sums whereas the expected values are subtracted. The expected values are given by the average of the previous window, see Equation 3. For estimating the sum of measurements and the sum of measurement squares within the window, our approach additionally introduces only two division, two addition and two subtraction operations per single processing. In comparison to the original parallel axis theorem, the calculation effort of our approach is $O(14)$. It also gets by with storing 3 numbers only, which are both sums and the size of the sliding window. Table 1 presents the efforts of discussed approaches in at nutshell.

$$\sigma_s = \sqrt{\frac{1}{s}\left(\sum_{i=n-s-1}^{n-1} x_i^2 + x_n^2 - \bar{x}^2\right) - \left(\frac{1}{s}\left(\sum_{i=n-s-1}^{n-1} x_i + x_n - \bar{x}\right)\right)^2}$$

$$(3)$$

3.4 How to Apply Estimated Significance of Changes

The variance of previous SNR readings (within the sliding window) provides the basis to give a statement about actual SNR readings. It enables to decide whether actual readings meet expected parameters or not. It further allows to assess how far new readings deviate from the expected scope. Therefore the system determines the significance indicator, which states by what multiple the actual reading diverges from the variance. Due to the parameters learnt from the variance of the measurements within the sliding window, the indicator automatically detects significant deviations without the need for predefined thresholds. That way, we expect enabling sensors to apply equal SNR processing for location-independent jamming detection.

4 Assessment of the Significance Analysis Approach

To give a proof of concept, we applied our approach on large data sets taken from our colleagues work presented in [4]. They set up a wireless network topology collecting Signal-to-Noise Ratio (SNR) measurements at different devices. Please note, their work originally focuses on general wireless networks instead of wireless sensor nodes. Used nodes were equipped with Mini ITX boards, with 512 MB RAM and a 80 GB hard disk. They are also equipped with Atheros NMP 8602 802.11 a/b/g wireless cards, controlled by the Madwifi MAC driver (version 0.9.4), on Ubuntu Linux.

4.1 Experimental Setup

Figure 1 represents the network topology used. The sender sent UDP traffic to the receiver at a constant rate of 18 Mbps. All measurements collected at the sender, the receiver and a monitor have been recorded for later processing. Sender, receiver and monitor all operate on channel 56. The network interface of the monitor was set to monitor mode, hence received all packets sent on this channel. Jamming is performed by using two further nodes, i.e., the evil sender and the evil receiver. These two nodes operated on channel 52, which is adjacent to channel 56 and hence, produces interference. The solid lines in

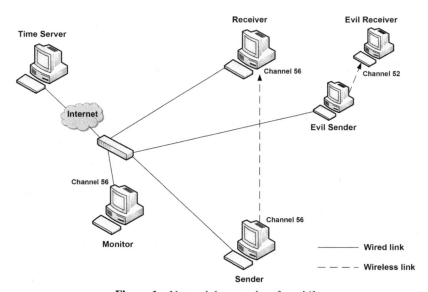

Figure 1   Network layout taken from [4]

Figure 1 represent a dedicated wired backbone for running the experiments. The evil sender periodically transmitted data with different inactive phases at a transmission power and rate of 13 dBm and 6 Mbps, respectively.

4.2 Evaluation Results

Figure 2(a) shows the impact of the evil communication on the SNR at the (not-jamming) sender and Figure 2(b) at the (not-jamming) receiver respectively. These figures also show short influences, from second 44.0 to 45.5 and long influences, from second 46.2 to 47.2. These result from the normal protocol behaviour between evil sender and evil receiver. Due to this behaviour jamming detection in wireless networks is challenging. Based on their measurements our colleagues successfully tested their intrusion, or jamming, detection algorithms. For further details please refer to their work in [4]. However, even though their approach provides good detection results, necessary calculation and memory effort of the algorithms used are far from being useable for WSN. Therefore, we tested our approach based on significance indication, which requires significantly lower effort, on the same data sets of SNR measurements. Figures 3(a) and 3(b) clearly show that our significance indication reliably detects all deviations in the SNR value. In addition, it reacts faster than the variance method, see the more detailed Figures 4(a) to 4(d) showing a short jamming period. This holds true on each falling edge of the SNR values independently of the position of the jamming device. From our point of view the detection speed is of high importance since

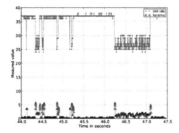
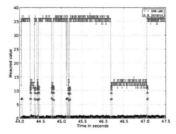

(a) SNR and respective variance at the sender.

(b) SNR and respective variance at the receiver.

Figure 2 Influence of jamming to Signal-to-Noise Ratio (SNR) at sender and receiver. It clearly shows the impact of jamming on SNR to vary according to the distance of the jammer. The SNR readings at the receiver, which is located in close distance to the jammer, considerably show larger drops than those at the sender. These drops consequently also lead to local amplitudes in the variance

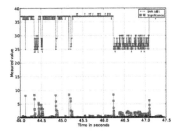

(a) Significance indicator results at the sender.

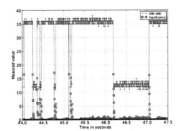

(b) Significance indicator results at the receiver.

Figure 3 Measured Signal-to-Noise Ratio (SNR) and results of the low-cost significance indicator approach based on SNR readings at sender and receiver

it can give the network under attack time to initiate proper counter measures. We are aware of the fact that there is not much a sensor node can do under jamming. But, storing sensed data, changing the wireless channel, trigger an alarm that jamming is ongoing are potential counter measures, and help to provide data for forensics and may be even a basis for a network manager to start counter measures, such as go and search the jamming device. On the one hand the high sensitivity of our significance approach is positive since it allows for detecting jamming from various positions within the WSN avoiding to set thresholds, which depend on the location of the jamming device. On the other hand triggering a jamming alarm whenever contention leads to a variation in the SNR value is unwanted since it leads to a significant number of false positives. In order to avoid false positives we are analysing the duration of the changes of the significance value. If it drops down to its initial value within the next values the change is not interpreted as jamming, since a jamming attack is considered to last longer. Our experiments figured out that drops in significance values lasting longer than ten percent of applied window size most probably identified a jamming period. Figures 4(a) to 4(d) display very short time slot with single or short drops in SNR with the associated variance and significance indication values. These figures clearly show that the duration of the significance value can be used as an indicator for jamming, due to its sensitivity. In case of jamming, i.e. second 45.16 from sender (Figure 4(a)), the significance indicator value stays high, whereas it drops down quickly in case of accidental increase of the SNR value. In contrast to this the variance method does not really allow to differentiate between accidental changes in the SNR value and intentional changes, i.e., jamming attacks. This is due to the fact that the variance analysis values do not change back quickly enough.

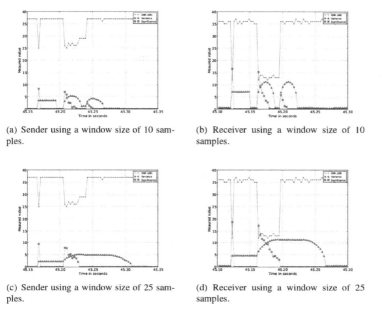

(a) Sender using a window size of 10 samples.

(b) Receiver using a window size of 10 samples.

(c) Sender using a window size of 25 samples.

(d) Receiver using a window size of 25 samples.

Figure 4 SNR, variance and significance indicator during single SNR drop and a short jamming period. Whereas the variance lags behind SNR changes, the significance indicator directly responds to each change

Figures 4(a) to 4(d) clearly show that they stay high even after a quick recovery of the SNR value.

A high significance indicator reflects the difference between the expected SNR value and the actually measured one. In other words, the significance indicator value is zero as soon as the measured value is close to the expected one. Since the expected SNR value changes slowly over time, i.e., it increases only gradually from measurement to measurement, the size of the sliding window influences the sensitivity of the significance indicator approach. In contrast to the immediate reaction of the significance indicator approach, the variance analysis only indicates changes in the expected SNR value. By that the indicated deviation between two measurements is much smaller. In addition with continuous change in SNR values, the expected SNR and the variance converge to the actual SNR value. When the variance has reached its local maximum, the change of the SNR is fully reflected by the values of sliding window determining the variance. Thereafter only new changes may be registered and the variance value moves slowly back.

4.3 Handling False Positives & False Negatives

The size of the sliding window examined to determine whether or not the jamming indication value drops back or not, has a significant impact on the speed with that the indication value drops down. For a window size of 10 samples the values of both approaches drop down to the initial value, but with different speed. The significance indicator drops down almost immediately whereas the variance analysis needs several milliseconds. If the window size is prolonged to 25 values, see Figures 4(c) and 4(d), both approaches react slower in case that the SNR value has changed for longer than single or few measurements. For improving the correctness of the significance indicator approach this behaviour is beneficial, since the value stays high longer than with shorter windows, i.e., it can be taken more seriously. For extremely short variations in the SNR value the significance indicator value still reacts immediately, in other words, there we do not loose accuracy.

There is another not yet reflected aspect when deciding whether changes in the SNR shall be interpreted as jamming or not. The type of MAC protocol is an important factor. In normal contention based MAC protocols our considerations are valid, but are they still true when the MAC protocol applies a certain type of schedule? In such environments even very short periods of interference (dips in the SNR) might be due to jamming. What we mean is that a sophisticated jammer could try to block a selected time slot, i.e., to interrupt the connection of a single device. As additional means to detect such situations, we propose to specify the type of MAC and to record in which MAC time slot variances of the SNR are recorded. If it becomes apparent that a specific time slot is more often affected than others, jamming might be the correct interpretation independently of the adaptation speed of the significance indicator value.

5 Significance Analysis Based Jamming Detection

To give a real proof of evidence of our approach, we certainly needed to implement the presented mathematical approach on a state of the art sensor node rather than using data measured in a laboratory network set-up only. Therefore all details and experimental results provided in this section result from implementing the significance analysis on our own sensor node platform presented in the next subsection.

5.1 Applied Sensor Node Platform

Our own sensor node platform used for implementing the jamming detection approach was originally developed for monitoring vital and environmental of fire-fighters in action (Piotrowski et al.2010). In this platform we have three different radio transceivers in parallel, these are:

- TI CC1101: Low cost transceivers working in the 868 MHz band
- TI CC2500: Low cost transceivers working in the 2.4 GHz band
- TI CC2520: ZigBee TM Transceiver working in 2.4 GHz band

We applied two pin and logic compatible transceiver chips working in different radio frequency bands, i.e. the first in the European 868 MHz band and the second in the 2.4 GHz. The third transceiver (ZigBee) is also working in the 2.4 GHz band and provides 802.15.4 support. It was applied to be able to communicate with other known node platforms. The sensor node is empowered by the MSP430F5438 Microcontroller from TI. The complete sensor node is depicted in Figure 5. Due to it's redundancy in radio connectivity, our platform provides the ideal basis to implement and test different jamming scenarios with one setting of nodes. All results presented in this section have been measured on this platform.

Figure 5 Top view of the applied sensor node platform in comparison to the size of a 1 Euro coin

5.2 Concept of Jamming Detection

The concept presented here is based on several components, which are depicted in Figure 6. RSSI values from the radio module (RF module) are used as input data. The first component of the jamming detection is a significance analysis

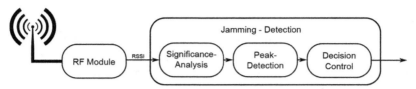

Figure 6   Concept of jamming detection

based on the mathematical approaches presented in the previous section 3.3. The result of this component are the significance changes in the RSSI values. We assume a large change in significance value to be caused by an active jammer. This information is used to generate an event to be evaluated by a following peak detection algorithm. In most peak detection algorithms a fixed threshold is used, but is not acceptable in this context. To decide between week and strong deflection, we use a dynamic threshold which is calculated with the Heaviside function given in equation 4.

$$\Theta(x)_{peak} = (\frac{1}{s} \sum_{i=n-s-1}^{n-1} x_i + x_n - \bar{x}) - (\frac{\max(x_n, x_{n-1}, ..., x_{n-s-1})}{2}), x \in \mathbb{N} \quad (4)$$

The last component of the jamming detection engine is the decision control, which finally signals whether jamming is ongoing or not. The flowchart of the engine is shown in Figure 7. First the engine checks whether an event was generated by the peak detection. If that is true then the old (avg_{old}) is compared to the new (avg_{new}) average value of the RSSI input data. This branch corresponds to the formula 5. The second branch is used to solve two problems. The first challenge is that a short interference phase determines only the start of the jamming activity but not the end. The second challenge occurs if a long interference phase exist. During this phase fluctuations in the RSSI average value can cause that the alarm is reset. To distinguish between short and long phase we use a counter and a fixed threshold (t_{phase}). The threshold depends on the size of the sliding window of significance s_{sign} and is calculated by $t_{phase}= 1.25\, s_{sign}$. In case merely a short phase of interference appears, the trend of RSSI values is checked, otherwise no further calculations are performed.

$$\Theta(x)_{decision} = ((\frac{1}{s} \sum_{i=n-s-2}^{n-2} x_i + x_n - \bar{x}) - (\frac{1}{s} \sum_{i=n-s-1}^{n-1} x_i + x_n - \bar{x})) \quad (5)$$

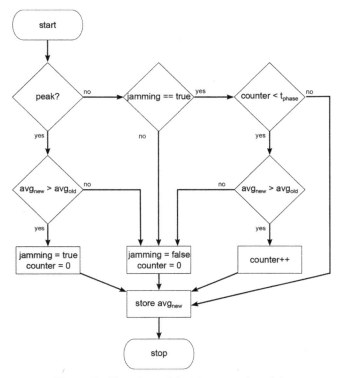

Figure 7 Flowchart of decision control module

5.3 Experimental Setup

The experimental setup is shown in Figure 8. We used two sensor nodes for a simple communication between sender and receiver and one node as a jammer. The position of the jammer varies between near sender (position 1), between sender and receiver (position 2) and near receiver (position 3). The data communication and the jamming activity is performed using the 2.4 GHz band on channel 15. To exchange control informations the 868 MHz band is used in parallel. For the experiment we implement two different application for the platform. The first application, called *CommApp*, periodically sends data packets with a data rate of 250 kbps and a payload of 60 Bytes from sender to receiver. The sending interval is 50 ms. Furthermore, the application periodically reads actual RSSI values from the CC2520 (2.4 GHz) transceiver at every 10 ms and forwards it as input for the jamming detection engine. To analyse the results we forward all important information in CSV format

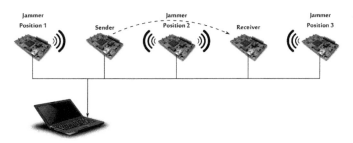

Figure 8   Experimental set-up of jamming scenarios and node positions

over the UART interface to a logging and control application running on the laptop. Collected data are the current RSSI value, intermediate data of jamming detection, packet delivery rate (PDR), jammer activity, number of received preambles and packets.

The second application, called *JammApp*, implements three types of jamming, i.e., constant-, periodic- and random jamming. To generate a disturbing signal with CC2520 radio we use a special transmit mode, which send pseudo random data [6]. The state of *JammApp* can be set to active or inactive and can toggle in periodic or random intervals. Before the activity of the jammer changes, the *JammApp* sends its state over the control channel via 868 MHz to all sensor nodes. This is needed to be able to check and synchronise the detection results with the actually attack state for evaluation purposes. The actual jamming scenario remains undisturbed by that.

The Figure 9 shows the impact of the communication and jammers on the wireless channel. The first graph (a) is a normal communication between sender and receiver without malicious impact. The graph (b) to (d) illustrate the impact of respective three different types of jamming. To test our detection algorithms we collected data for each scenario over a 30 min time period. For all scenarios following parameters are used:

Clock rate MSP430: 4 MHz

Sample rate: 10 ms

Sliding window of significance analysis: $s_{sign} = 8$

Sliding window of peak detection: $s_{peak} = 2$

5.4 Implementation Details

Resources used

An overview of the memory consumption is given in Table 2. The table illustrates the size of code (text) and the required memory space (data) of the jamming detection mechanism. The individual functions have been

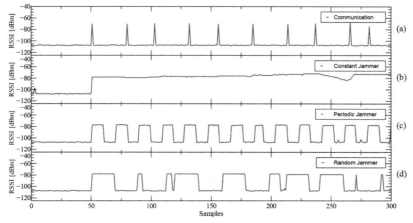

Figure 9 Monitored RSSI values of communication (a), constant jammer (b), periodic jammer (c), random jammer (d)

Table 2 Static analysis - memory requirements

Function	Memory in [Byte]	
	Text	Data
Significance Init() Significance Eval()	33 326	16
PeakDetection Init() PeakDetection Eval()	22 198	17
DecisionControl Init() DecisionControl Eval()	18 358	14
Total	955	47

disassembled from the binary files to determine the amount of memory needed. The total text size of the detection engine is 955 bytes only, which is quite acceptable also for sensor nodes. The biggest amount of memory is used for the decision control, which is due to the additionally implemented mechanisms as mentioned above. The space required for the storing values for calculations during runtime is even much lower. We here need 47 bytes for variables only. This gives a total of about 1 KB (1002 bytes) of required memory, which is in a tolerable range even for the limited memory resources of sensor nodes.

Online jamming detection tool

To configure the test scenarios we implemented two applications. Both used the virtual serial com port to communicate with the sensor node platform to actively control the scenarios during run-time. The first application, shown on the left side of Figure 10, allows us to set the wireless channel and the transmit

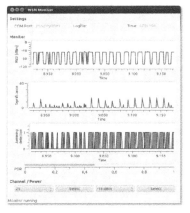

Figure 10   Graphical user interface of remote interface from sensor node (left) and jammer (right) application

power of the radio transceiver of connected sensor node. Furthermore it allows for plotting and storing the current values of detection analysis in real-time. The second application is build for the configuration of the *JammApp*. Here, the type and activity of the jammer can be set in addition to the pure radio configuration.

Runtime

The time needed to perform the algorithms was measured by gathering required clock cycles. The interrupts of the used timer were configured such that a cycle corresponds to one microsecond in time, at a processor clock rate of 4 MHz. The results are shown in Table 3. A single detection run may require up to 4944 cycles. Thus, a single detection cycle runs approximately 4.9 ms at a clock rate of 4 MHz. If the constant jammer is active, the required time decreases to 4.2 ms (4237 cycles) only. This is due to less processing of data in the decision control, since most of the decision control functionality is only activated when significant changes are detected, i.e. the constant jammer is the best to be recognized very quickly.

Experimental results and detection rates

The most crucial goal of each jamming detection method is the effectiveness and accuracy of detecting attacks correctly. Therefore we have separated the number of active jamming phases in relation to the correctly detected jamming phases from the measured data.

Table 3 Captured data and accuracy of jamming detection applying various positions of the jamming device

Criteria	Node	Jammer position 1 (near Sender)		
		Constant	Periodic	Random
RSSI in dBm	Receiver	−88.36	−97.39	−98.1
	Sender	−63.29	−84.17	−84.66
runtime in clock cycles	Receiver	4423	4727	4716
	Sender	4361	4941	4812
Detected attacks in %	Receiver	100	65.72	53.9
	Sender	100	99.67	89.09
PDR in %	Receiver	97.08	99.16	98.94
Criteria	Node	Jammer position 2 (in between)		
		Constant	Periodic	Random
RSSI in dBm	Receiver	−82.86	−91.11	−91.17
	Sender	−66.79	−86.19	−86.66
runtime in clock cycles	Receiver	4287.3	4916.91	4806.31
	Sender	4243.69	4944.53	4811.25
Detected attacks in %	Receiver	100	99.92	90.06
	Sender	100	99.82	87.62
PDR in %	Receiver	1.01	47.58	46.78
Criteria	Node	Jammer position 3 (near receiver)		
		Constant	Periodic	Random
RSSI in dBm	Receiver	−65.17	−82.24	−83.11
	Sender	−78.42	−94.97	−95.52
runtime in clock cycles	Receiver	4237	4928	4820
	Sender	4289	4779	4728
Detected attacks in %	Receiver	100	99.98	91.19
	Sender	100	91.89	84.47
PDR in %	Receiver	0.42	47.47	46.54

First, the attacks of the constant jammer are evaluated. The detection rates for all scenarios with a constant jammer are 100 %. Sender and receiver have thus achieved the same results, although different distances have been tested, i.e., the different positions of the jamming device. That outcome slightly changes when considering the results while the periodic jammer was active. Here not all activities have been recognized correctly and the results are coupled with the distance to the jamming device, as pre-assumed before testing. If the jammer is located in-between sender and receiver at the same distance to both sensor nodes (position 2), then both nodes detect almost all attacks. The maximum measured error rate was 0.18 %. Higher error rates can only be seen at the positions 1 (close to the transmitter) and 3 (near the receiver). At position 1 there is a error rate in contrast to position 3.

The reason for this are low deviant RSSI values and due to the simple fact, that transmission is disturbed at the sending device. Thus, a packet may be destroyed while sending already or more important, the sender does not send at all due to already occupied channel. Most MAC protocols of course sample the wireless channel before sending. Nevertheless detection rates of 65.7 % with a minimal increase in the RSSI average of 8.9 dBm can still be obtained. Better results have been measured at position 3. In this scenario, we achieved detection rates of 99.98 % at the receiver and of 91.89 % at the sender respectively. The most difficult jammer to be detected was the random jammer. The rapid change from short to long interfering signals are reflected in the results again. It still reaches values of 84 % to 93.4 %. However, the lower detection results at the receiver and the jammer at position 1 still remain due to low RSSI values returned. Overall, very good results have been achieved in all scenarios given to the fact that not pre-configured jamming is difficult at all, especially in sensor networks. It has been shown that even weak jammers were effectively detected. In a sensor network scenario, all nodes located near to the jammer should be able to detect an attack. Hence, also nodes not affected by the jammer can then inform all other nodes as well as a sink about on-going jamming and take any necessary corrective action, e.g. searching for a suitable other wireless channel.

6 Concluding Remarks

In this paper we have introduced the significance analysis approach as a means to detect jamming in wireless sensor networks in real-time. Since it reacts on changes in the monitored value, here SNR, our approach omits the need for preconfigured values, which are difficult to get and in addition are difficult to use since the impact on the SNR depends on the position of the jammer. We have evaluated the complexity of our approach and shown that in theory the computational effort equals to O(14) and that the memory consumption collapses to storing three integer values. We have evaluated the correctness of the prediction by applying it to real measurements recorded in a jamming experiment.

To provide evidence, we have implemented and tested our jamming detection engine based on the significance analysis on a state of the art sensor network setting. It requires about one kilobyte of code and data memory only. Our jamming detection engine monitors the wireless channel with a sampling rate of 10 ms. It returns a jamming detection decision within less than 5 ms and achieves a detection accuracy in between 84 to 99 percent. It turned out that the obviously best position of jammer is near the receiver, since then the jamming might be overseen by the sending device whereas the receiving one is not able to send an alert message due to ongoing jamming either.

Even though we have proven our approach to be working quite well, there still exist several open questions such as correlation of detection method and used MAC protocols, reaction time to be gained e.g. in case the jamming device slowly increases its transmission power etc. Especially the latter case is of interest for us, since in this case jamming might probably not be detected as a peak in SNR by our approach. In summary, jamming detection methods in general need to be seen as continuous work in progress due to the simple fact that jamming methods and jamming devices will advance too.

References

[1] *Wireless Communications: Principles and Practice.* Prentice Hall communications engineering and emerging technologies series. Dorling Kindersley, 2009.

[2] J.B.D. Cabrera, C. Gutierrez, and R.K. Mehra. Infrastructures and algorithms for distributed anomaly-based intrusion detection in mobile ad-hoc networks. In *Military Communications Conference, MILCOM 2005. IEEE*, pages 1831–1837. IEEE, 2006.

[3] Murat Çkiroğlu and Ahmet Turan Özcerit. Jamming detection mechanisms for wireless sensor networks. In *Proceedings of the 3rd international conference on Scalable information systems*, InfoScale '08, pages 4: 1–4: 8, ICST, Brussels, Belgium, 2008. ICST.

[4] A. Fragkiadakis, V. Siris, and N. Petroulakis. Anomaly-based intrusion detection algorithms for wireless networks. In *8th International Conference on Wired/Wireless Internet Communications*, pages 192–203, Lulea, Sweden, June 2010. Springer.

[5] A. Hamieh and J. Ben-Othman. Detection of jamming attacks in wireless ad hoc networks using error distribution. In *Communications, 2009. ICC '09. IEEE International Conference on*, pages 1–6, June 2009.

[6] Texas Instruments. *CC2520 DATASHEET –2.4 GHZ IEEE 802.15.4/ZIGBEE RF TRANSCEIVER.*

[7] Yu Seung Kim, Frank Mokaya, Eric Chen, and Patrick Tague. All your jammers belong to uslocalization of wireless sensors under jamming attack. In *Communications (ICC), 2012 IEEE International Conference on*, pages 949–954. IEEE, 2012.

[8] Yee Wei Law, Pieter Hartel, Jerry Hartog den, and Paul Havinga. Link-layer jamming attacks on s-mac. In *Proceeedings of the Second European Workshop on Wireless Sensor Networks, 2005*, pages 217–225, Los Alamitos, California, 2005. IEEE Computer Society Press.

[9] R. Muraleedharan and L.A. Osadciw. Jamming attack detection and countermeasures in wireless sensor network using ant system. *SPIE Defence and Security, Orlando*, 2006.

[10] Alejandro Proano and Loukas Lazos. Selective jamming attacks in wireless networks. In *Communications (ICC), 2010 IEEE International Conference on*, pages 1–6. IEEE, 2010.

[11] K.W. Reese, A. Salem, and G. Dimitoglou. Using standard deviation in signal strength detection to determine jamming in wireless networks. *Computers Applications in Industry and Engineering*, 2010.

[12] Mario Strasser, Boris Danev, and Srdjan Čapkun. Detection of reactive jamming in sensor networks. *ACM Transactions on Sensor Networks (TOSN)*, 7:16:1–16:29, September 2010.

[13] J. Tang and P. Fan. A RSSI-based cooperative anomaly detection scheme for wireless sensor networks. In *International Conference on Wireless Communications, Networking and Mobile Computing, WiCom 2007.*, pages 2783–2786, Shanghai, 2007.

[14] A. D. Wood, J. A. Stankovic, and Gang Zhou. DEEJAM: Defeating Energy-Efficient Jamming in IEEE 802.15.4-based Wireless Networks. In *4th Annual IEEE Communications Society Conference on Sensor, Mesh and Ad Hoc Communications and Networks, SECON '07.*, pages 60–69, 2007.

[15] Wenyuan Xu, Ke Ma, W. Trappe, and Y. Zhang. Jamming sensor networks: attack and defense strategies. *Network, IEEE*, 20(3): 41–47, May 2006.

[16] Wenyuan Xu, Wade Trappe, Yanyong Zhang, and Timothy Wood. The feasibility of launching and detecting jamming attacks in wireless networks. In *Proceedings of the 6th ACM international symposium on Mobile ad hoc networking and computing*, MobiHoc, pages 46–57, NY, USA, 2005.

Biographies

Stephan Kornemann received the M.S. degree in information and media technology from Brandenburg University of Technology, Germany, in 2012. From 2010 to 2012 he worked as software tester at Philotech and received his software testing qualification certificate from ISTQB. Since 2012 he is member of the sensor network research group at IHP in Frankfurt (Oder). His current research focuses on intrusion detection systems for wireless sensor networks.

Steffen Ortmann received his diploma in computer science in 2007 and his PhD in engineering by scholarship in 2010. Since 2005 he is active in the sensor network research group of IHP. He has published more than 40 refereed technical articles about reliability, privacy and efficient data processing in wireless sensor networks and medical applications. His current research focuses on mobile wireless sensor networks for tele-medical inno-vations. He is coordinating the FP7 project StrokeBack and is responsible for medical driven research and the crypto-microcontroller research team within sensor networks group of IHP.

Peter Langendörfer, Professor, holds a diploma and a doctorate degree in computer science. Since 2000 he is with the IHP in Frankfurt (Oder). There, he is leading the sensor networks and mobile middleware group. Since 2012 he has his own chair for security in pervasive systems at the Technical University of Cottbus. He has published more than 100 refereed technical articles, filed ten patents in the security/privacy area and worked as guest editor for many renowned journals e.g. Wireless Communications and Mobile Computing (Wiley). He was chairing International conferences such as WWIC and has served in many TPC for example at Globecom, VTC, ICC and SECON. His research interests include wireless sensor networks and cyber physical systems, especially privacy and security issues.

Alexandros Fragkiadakis received his PhD degree in computer networks from the Department of Electronic and Electrical Engineering of Loughborough University in UK. His thesis focused on active networks using programmable hardware (FPGAs). He has also received an MSc in Digital Communications Systems, awarded with distinction, from the same University. Alexandros obtained his Diploma degree in Electronics from the Technological Educational Institute of Piraeus. He has worked as a Research Associate within the High Speed Networks Group of Loughborough University, in a project involving active networks and field programmable gate arrays. This project was funded by the Engineering and Physical Sciences Research Council, a British Government's leading funding agency for research and training in engineering and the physical sciences. Alexandros joined the Telecommunications and Networks Laboratory of the Institute of Computer Science of the Foundation for Research and Technology-Hellas in November 2008. His research interests include wireless networks, intrusion and anomaly detection in wireless networks, reprogrammable devices, cloud computing, open source architectures, cognitive radio networks, wireless sensor networks.

An Analysis of DoS Attack Strategies Against the LTE RAN

Jill Jermyn[1], Gabriel Salles-Loustau[2], and Saman Zonouz[2]

[1] *Department of Computer Science, Columbia University New York, NY*
jill@cs.columbia.edu
[2] *Department of Electrical and Computer Engineering, University of Miami,*
Miami, FL g.sallesloustau@umiami.edu, s.zonouz@miami.edu

Received 1 March 2014; Accepted 15 April 2014;
Publication 2 July 2014

Abstract

Long Term Evolution (LTE) is the latest 3GPP mobile network standard, offering an all-IP network with higher efficiency and up to ten times the data rates of its predecessors. Due to an increase in cyber crime and the proliferation of mobile computing, attacks stemming from mobile devices are becoming more frequent and complex. Mobile malware can create smart-phone botnets in which a large number of mobile devices conspire to perform malicious activities on the cellular network. It has been shown that such botnets can cause a denial of service (DoS) by exhausting user traffic capacity over the air interface. Through simulation and with studies in a real-world deployment, this paper examines the impact of a botnet of devices seeking to attack the LTE network using different types of strategies. We quantify the adverse effects on legitimate users as the size of the botnet scales up in both sparsely and densely-populated cells for varying traffic Quality of Service (QoS) requirements. Our results show that a single attacker can drastically reduce the QoS of legitimate devices in the same cell. Furthermore, we prove that the impact of the attack can be optimized by tuning the attack strategy, leveraging the LTE uplink MAC scheduler.

Keywords: LTE, DoS, security, mobile malware, botnets.

Journal of Cyber Security, Vol. 3 No. 2, 159–180.
doi: 10.13052/jcsm2245-1439.323

1 Introduction

Smartphones are becoming increasingly popular as multipurpose portable computing devices that run a complete software stack from the operating system to the user-level applications. Based on reports by ComScore Inc. [30], 110 million Americans used smartphones in 2012, and smartphones constitute 47% of the total mobile communication devices. Smartphone applications serve various sensitive and critical functionalities. At the same time, they are often developed by possibly untrusted and inexperienced third-party developers that may introduce new attack vectors and exploitable vulnerabilities. The increasing popularity of smartphones along with emerging possible attack vectors and vulnerabilities has turned them into appealing targets for malicious adversarial parties. Real-world solutions need to be designed to provide smartphone platforms with effective and practical security services. However, existing smartphone platforms generally face computational and storage limitations that hinder permanent deployment of comprehensive and heavyweight intrusion prevention and detection solutions.

As shown by many past research projects, smartphones remain vulnerable to various exploitation techniques. Intrusions on smartphones occur in one or more forms of the following three categories [14]. First, adversaries may violate user data confidentiality by exfiltrating device-resident, sensitive data to malicious end points. Second, attackers may cause data integrity violation via malicious unauthorized data modification within the user smartphones. Finally, the adversaries may target the availability of the system services provided by the smartphone platform. In this paper, we concentrate on infected devices targeting availability in the cellular network.

Long Term Evolution (LTE) is the latest cellular network standard for high-speed mobile devices. In 2013 there were 200 million devices connected over LTE [3], and this number is expected to surge to 1 billion by 2016 [22]. Not only are people reliant on LTE for their voice and data services, but with the rise of the Internet of Things (IoT), LTE has become an important resource for Machine-to-Machine (M2M) communication. According to Gartner there will be 26 billion devices on the IoT by 2020 [1]. Clearly, any impact on the availability of LTE services could cause catastrophic repercussions for the great number of devices that rely on them.

In this paper, we introduce a new set of denial of service (DoS) attack strategies that handheld devices could potentially use against LTE technologies. Malware executing these strategies would be neither particularly designed against a specific executable nor architectured against generic computing devices. Instead, it would make use of how the LTE access networks manage a large-scale networked system consisting of thousands of end-user devices. Consequently, all devices utilizing LTE solutions become potentially vulnerable. Due to the frequency spectrums available to LTE networks, there exists a limited capacity that the air interface can handle, making the whole infrastructure vulnerable to physical layer DoS intrusions.

We investigate uplink data traffic and examine how the LTE Medium Access Control (MAC) scheduler allocates resources for devices during both normal benign and adversarial scenarios. Furthermore, we determine the optimal size of a botnet in a single cell that is needed to significantly hinder availability of service/degrade legitimate customer Quality of Service (QoS).

We have implemented and evaluated a real working prototype of our proposed attack strategies on the most recent Android platforms. Our experimental results empirically prove the feasibility of the attacks against LTE networks. They also show that the usability of smartphone devices could be significantly affected through installation of malicious legitimate-looking applications on the device. Our simulation results from a large-scale network show that deployment of these strategies could potentially impact the data network efficiency for all legitimate end-user devices sharing the same cell.

In summary, the contributions of this paper are the following:

- We introduce novel LTE-specific denial of service attack strategies against smartphone devices that make use of LTE networks. The attacks are not detectable by traditional signature-based detection solutions.
- We demonstrate the attack effectiveness through simulation and determine the optimal botnet size and type of flooding traffic that would be necessary to severely impact legitimate users in the same cell.
- We implement a working prototype for the most recent Android platform and evaluate its efficiency and practical deployability in a real-world setting.

This paper is organized as follows. Section 2 describes the related literature and real-world intrusions against resource-limited smartphone devices, followed by an explanation of how cellular botnets can be formed and a discussion on instances of them in the wild. Section 3 gives a high-level background on state-of-the-art LTE access networks and describes uplink scheduling in the LTE Radio Access Network (RAN). Our simulation experimental setup and results from a set of attack strategies are presented in Section 4, whereas Section 5 talks about practical applications and implementation in a real-world setting. Finally, Section 6 concludes the paper.

2 Related Work

In this section we review the past work on adversarial remote hacking techniques against smartphones as well as smartphone-originating attacks against the cellular and data network infrastructures.

Mylonas et al. [27] perform a fairly thorough empirical proof of how feasible and simple it is for average programmers to develop smartphone specific malware. The authors continue with an in-depth investigation of the deployed security mechanisms within various smartphone platforms and conduct a comparison among those platforms. A similar study was conducted by Jeon et al. [18], who performed a security analysis of smartphones and proposed potential countermeasures. Marforio et al. [24] design a new covert channel and application collusion attack that was previously unknown against the Android platforms. Using their proposed attack vector, the attackers could penetrate into the smartphone devices and steal sensitive user data through development of several individually legitimate-looking applications that co-operatively gain a sufficient set of permissions for user privacy violation. Aviv et al. [8] design and develop a novel intrusion against handheld tools that make use of touch screens using the oily residues on those screens. The authors show that it is possible to infer the user passwords if he/she enters it using the touch screen.

There have also been attacks proposed that could originate from smartphone devices against the cellular and data network infrastructures. Ricciato et al. [31] review potential and previously proposed denial of service attack models specifically designed against cellular and data networks. The authors discuss the trade-offs between optimality and robustness of the designed cellular networks. They determine that a single attacker can create traffic profiles that are capable of straining the entire network infrastructure. Traynor et al. [34] characterize the impact of the large-scale compromise and coordination of mobile phones in attacks against the core of cellular networks. The authors demonstrate that a botnet comprised of 11K phones could potentially degrade service to area-code sized regions by 93%. In another similar effort, Enck et al. [13] empirically show the feasibility of SMS-based attacks against the cellular networks to induce a denial of voice service to cities the size of Washington D.C. and Manhattan.

2.1 Botnet Threats in Cellular Networks

The vast expansion and increasing popularity of highly-capable but largely insecure smartphone devices that are often interconnected with the Internet is a significant threat against large-scale cellular networks. In this section, we describe the well-known botnet intrusions in such cellular infrastructures and review the results from the past related literature.

Generally, botnets are defined as a collection of Internet-connected compromised computing devices, so-called zombies, communicating with other compromised systems in order to perform (often malicious) tasks. Unlike network worms, zombies are not autonomous and need to be ordered regarding what to do at each time instant. Such control orders may come from various communication channels such as an Internet Relay Chat (IRC) channel where the master sends control commands to be executed by the distributed zombies. Of the typical botnet tasks, one could mention large-scale spamming where all the zombies are ordered to send spam emails to the same target address, potentially causing a denial of service. Other similar intrusions may target Internet core services such as DNS [11] and BGP [21]. Traditionally, botnets have targeted desktop computer systems; however, with the increasing popularity of vulnerable and capable smartphone devices, the number of smartphone-specific botnets have recently risen by a significant factor.

In particular, smartphone botnets could potentially be more damaging to the underlying network infrastructures because cellular networks have more rigid hierarchical dependencies and hence are less likely to withstand similar misbehaviors. The past academic efforts investigating the possibility of a large-scale botnet attack against cellular networks have mainly consisted of two major categories [34]. First, researchers have attempted to explore whether the lack of authentication for signaling traffic in the wired network would enable an attacker with a physical connection to cause significant damage [20]. Second, there have been efforts to determine whether the same amount of damage is feasible by gathering a large set of compromised wireless devices and trying to either saturate the cellular network [35] or make use of the compromised smartphones as a spam generator to attack Internetbased resources [26]. The authors of [25] show that the threats described in [34] are concrete, and they demonstrate the ease of creating a mobile botnet on popular smartphones models.

There have been several instances of mobile botnets seen in the wild, most recently targeting the Android platform. MisoSMS, uncovered in December 2013, is one of the largest mobile botnets yet seen, stealing SMS messages and emailing them to a command-and-control infrastructure located in China [15]. One of the first Android malware to exhibit botnet-like capabilities was Geinimi, discovered in 2011, where a remote server had the ability to control and send instructions to infected devices [17]. However, Android is not the only platform vulnerable to such malware. iKee.B, released throughout several European countries in 2009, spread to jailbroken iPhones by using the default SSH password. The malware stole sensitive information and performed malicious activities on infected devices by controlling them through a Lithuanian botnet server [28].

On a more comprehensive investigation of the smartphone botnets, Traynor et al. [34] evaluated the impact of the large-scale infection and co-ordination of mobile phones in attacks against the core of a cellular network, namely denial of service attacks using selected service requests on the central repository of user location and profile information in the network. According to their results, such attacks could degrade the core services even in cellular networks with capable databases to the extent of approximately 75% when the size of the botnet reaches around 140K zombies.

Cambiaso et al. [10] present a comprehensive survey of DoS intrusions targeting the data service network. The authors classify those intrusions into two main categories. First, attacks that involve high-bandwidth, flood-based approaches exploiting vulnerabilities of networking and transport protocol layers. Second, slow-rate attacks that exploit vulnerabilities of application layer protocols to accomplish DoS objectives. Specifically, there have been several real-world and experimental malware samples performing denial of service intrusions. Dondyk [12] presents a DoS attack against smartphones that prevent non-technical smartphone users from utilizing data services by exploiting the connectivity management protocol when encountered with a WiFi access point. Gobbo et al. [16] describe a DoS attack against the data provider network via an unauthenticated injection of malicious traffic in the mobile operator's infrastructure that causes significant service degradation.

[29] exploits opportunistic scheduling in 3G networks by showing that devices can report false channel condition reports. The authors demonstrate that only five malicious devices in a 50-user cell can consume up to 95% of timeslots, in effect causing 2 second end-to-end packet delay on VoIP applications that make them virtually useless. There has been some recent research on DoS attacks against the LTE air interface. For example, the authors of [19] use simulations to determine the number of attackers needed to degrade service of legitimate VoIP users. Although peripherally similar to our work in scope, this research does not examine the impact of a botnet due to various QoS requirements nor does it provide results from a real-world environment.

Bassil et al. [9] simulate an attack against LTE where malicious devices request high-bandwidth GBR bearers while having a low Modulation and Coding Scheme (MCS) index. They demonstrate denial of service for legitimate TCP-based applications with two malicious devices and show that high priority voice bearers are not affected by attacks since they preempt video bearers. Their results are insightful and support many of our claims in this paper, yet their experiments do not show results for variable sized botnets nor a variable number of legitimate users. Additionally, the authors only consider scenarios where malicious devices request bandwidth for high QoS applications, such as conversational video, and fail to demonstrate that botnets demanding low QoS applications can severely impact resource availability as well. In our paper, however, we quantify the impact of a botnet on legitimate users as the number of malicious devices scales for both lightly-used and densely-populated cells, and for both low QoS and high QoS traffic.

3 LTE Radio Access Network

The LTE radio interface, also known as the Uu interface, lies between the eNodeB and the User Equipment (UE). Offering high peak transmission rates, the physical layer implements Orthogonal Frequency-Division Multiple Access (OFDMA) in the downlink and Single-Carrier FDMA (SC-FDMA) in the uplink. Data and Signaling Radio Bearers (DRB and SRB) transmit user-plane and control-plane traffic on the air interface. DRBs support many different types of service requirements, for example voice call and mobile broadband access, through their QoS configuration [32]. QoS describes the combination of requirements for categories of data traffic, including latency, guaranteed bit rate (GBR), jitter, and error rate. Classes of traffic can be grouped by a QoS Class Identifier (QCI) which indicates the traffic's requirements for delay, priority, and error rate. Figure 1 shows a list of standard QCI profiles as defined by 3GPP [7]. End-to-end QoS is important for distinguishing certain real-time services, such as voice, from basic broadband access and providing them with reliable delivery.

QoS scheduling over the air interface in LTE in both the uplink and downlink direction is handled by the MAC scheduler in the eNodeB [23]. Non-GBR bearers, which do not guarantee minimum resources, are allocated bandwidth depending on the QCI of the service and the current cell utilization. They are typically provided with service according to fairness criteria. Examples of applications using Non-GBR bearers are web browsing and FTP. GBR bearers, on the other hand, are granted throughput up to an agreed-on guaranteed bit rate depending on the current cell utilization and in some cases may force a lower priority user to relinquish services. Real time applications, voice, and streaming video typically use GBR bearers. The job

QCI	Resource Type	Priority	Packet Delay Budget	Packet Error Loss Rate	Example Services
1	GBR	2	100 ms	10^{-2}	Conversational Voice
2	GBR	4	150 ms	10^{-3}	Conversational Video (Live Streaming)
3		3	50 ms	10^{-3}	Real Time Gaming
4		5	300 ms	10^{-6}	Non-Conversational Video (Buffered Streaming)
5		1	100 ms	10^{-6}	IMS Signalling
6		6	300 ms	10^{-6}	Video (Buffered Streaming) TCP-based (e.g., www, e-mail, chat, ftp, p2p file sharing, progressive video, etc.)
7	Non-GBR	7	100 ms	10^{-3}	Voice, Video (Live Streaming) Interactive Gaming
8		8	300 ms	10^{-6}	Video (Buffered Streaming) TCP-based (e.g., www, e-mail, chat, ftp, p2p file
9		9			sharing, progressive video, etc.)

Figure 1　3GPP standardized QCI Characteristics

of the MAC scheduler is to allocate the air interface resources so that bearer QoS requirements are met and priorities among different QCIs are sustained [6]. Although the exact algorithms used in scheduler implementation are vendor specific, they must balance fairness and QoS when making allocation decisions.

Studying uplink scheduling related to LTE network availability is particularly important with the rise of M2M communications using 4G networks, as most M2M applications are uplink dominant and will therefore make high demands on uplink bandwidth [33]. The manner in which the MAC scheduler handles such voluminous requests for uplink resources will play an important role in how customers are affected by high service demands, whether they be of legitimate or malicious origin. In addition, QoS distinctions in LTE will become increasingly important for M2M applications that have strict requirements for delay and reliability of service, such as medical devices and smart grid.

4 Simulation Experiments and Results

This section describes our set of simulation experiments for testing multiple DoS attack strategies against the LTE RAN. As described in Section 3, the LTE MAC scheduler is responsible for making bandwidth allocation decisions and is influenced by the particular traffic QoS requested for a bearer. Our experiments illustrate that these scheduling decisions can be exploited to optimize certain DoS attack strategies. We show that the network fails to recognize a botnet as malicious and consequently tries to furnish it with resources by reducing legitimate device throughput.

4.1 Experimental Setup

All of our experiments were performed on the OPNET Modeler's LTE model [4], a comprehensive platform that is standards compliant at Release 8. The model consists of several network elements, including the UEs, eNodeB (LTE base station), Evolved Packet Core (EPC), and IP servers. Traffic sent from UEs is customizable with regard to QoS, intensity, start time, and duration. For our experiments, we selected a 3MHz LTE deployment with a single cell and designated a subset of UEs as attackers that attempt to saturate the RAN with large amounts of uplink traffic at a particular time during the simulation. The remaining UEs depict legitimate devices, as their traffic profiles are set within reasonable typical usage patterns. Figure 2 shows

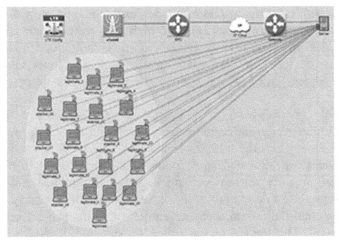

Figure 2   Simulation screenshot of legitimate and malicious UEs in a single LTE cell

a screenshot of an OPNET simulation scenario containing both legitimate and malicious UEs in a single cell.

We simulated several attack scenarios to determine the impact of various types of botnets on legitimate user QoS. In each experiment we modified the size of the botnet, the number of legitimate users in the cell, and the QoS requirements of the traffic sent by both malicious and legitimate devices. Video conferencing is indicated by traffic with QCI 2, while QCI 9 traffic represents web browsing or file transfer.

4.2 Simulation Experiments and Results

Our first experiment deployed a lightly-used cell of 20 legitimate users each sending traffic to a remote IP server at 100,000 b/s for 20 minutes. During the traffic period a botnet of malicious devices attempts to flood the network by sending 2 Mbps per device of uplink traffic. Table 1 (a) and (b) summarize the impact of the botnet on the cell's legitimate devices for an increasing number of malicious devices and for traffic with varying QCI. Clearly the impact of the attack fluctuates based on the QCI of the traffic sent by both the legitimate and malicious devices. In general, when legitimate devices send traffic with a low QCI, a botnet can cause a complete denial of service for those devices. For example, Table 1 (a) shows that a botnet of only one device precipitates zero throughput for 80% of the legitimate users. We see a similar impact even when the attacker's traffic has a high QCI. The reason for the complete rejection of allocated bandwidth is that the scheduler is unable to furnish the guaranteed bit rate of the legitimate UEs and therefore denies them completely. It is also interesting to see that when attackers send high QCI traffic, increasing the botnet size by 20 times doesn't exacerbate the impact any more than with a single malicious device. However, when the attackers send low QCI traffic, we see a further deterioration of legitimate throughput when the botnet reaches 15 devices.

When legitimate traffic has a high QCI, shown in Table 1 (b), the botnet attack inflicts different consequences. We see a drastic reduction in legitimate UE QoS when the attackers send low QCI traffic. Although all legitimate devices are allocated some uplink bandwidth, their throughput declines so significantly during the attack that their applications would be virtually unusable. For example, the throughput of legitimate users is reduced to less than half when there are only two malicious devices sharing the same cell. The impact of the attack proves significantly worse as the size of the botnet increases. Table 1 (b) shows that when the botnet

Table 1 Uplink throughput for legitimate UEs in a lightly-used cell for different QoS traffic and varying botnet size, (a) Legitimate traffic: QCI 2, (b) Legitimate traffic: QCI 9

Size of Botnet [number of devices]	Attacker Traffic: QCI 9 [percentage of legitimate devices: uplink throughput]	Attacker Traffic: QCI 2
1	20%: 100,000 b/s 80%: 0 b/s	20%: 100,000 b/s 80 %: 0 b/s
2	20%: 100,000 b/s 80%: 0 b/s	20%: 100,000 b/s 80 %: 0 b/s
5	20%: 100,000 b/s 80%: 0 b/s	20%: 100,000 b/s 80 %: 0 b/s
10	20%: 100,000 b/s 80%: 0 b/s	20%: 100,000 b/s 80 %: 0 b/s
15	20%: 100,000 b/s 80%: 0 b/s	10%: 100,000 b/s 10 %: 28,000 b/s 60%: 0 b/s
20	20%: 100,000 b/s 80%: 0 b/s	5%: 31,000 b/s 15 %: 13,000 b/s 80%: 0 b/s

(a)

Size of Botnet [number of devices]	Attacker Traffic: QCI 9 [legitimate uplink throughput]	Attacker Traffic: QCI 2
1	71,500 b/s	62,500 b/s
2	65,700 b/s	43,000 b/s
5	58,000 b/s	15,000 b/s
10	46,000 b/s	4,100 b/s
15	40,000 b/s	190 b/s
20	35,300 b/s	190 b/s

(b)

scales from 1 to 5 devices, the bit rate of each legitimate user is reduced by 76%. When it reaches 15 malicious devices, the legitimate throughput becomes less than 0.2% the traffic rate actually sent by each legitimate device.

Our second set of simulation experiments sought to disclose the effect of a botnet on a single legitimate device that attempts to transmit data during an ongoing attack, again for a lightly-used cell of 20 legitimate devices. The device sends 200,000 b/s uplink traffic once all malicious devices initiate the attack (the remaining legitimate devices still send 100,000 b/s uplink traffic each). Table 2 summarizes the results of these experiments. In all cases where the attackers send low QCI traffic, the single legitimate UE is completely

Table 2   Uplink throughput for a legitimate UE in a lightly-used cell that attempts to connect during an ongoing attack for different QoS traffic and varying botnet size, (a) Legitimate traffic: QCI 2, (b) Legitimate traffic: QCI 9

Size of Botnet [number of devices]	Attacker Traffic: QCI 9	Attacker Traffic: QCI 2 [legitimate uplink throughput]
1	0 b/s	0 b/s
2	0 b/s	0 b/s
5	0 b/s	0 b/s
10	0 b/s	0 b/s
15	0 b/s	0 b/s
20	0 b/s	0 b/s

(a)

Size of Botnet [number of devices]	Attacker Traffic: QCI 9	Attacker Traffic: QCI 2 [legitimate uplink throughput]
1	120,000 b/s	96,000 b/s
2	120,000 b/s	71,000 b/s
5	66,000 b/s	20,000 b/s
10	46,000 b/s	4,000 b/s
15	48,000 b/s	190 b/s
20	40,890 b/s	190 b/s

(b)

denied bandwidth, even with only one attacker in the cell. When sending traffic requiring high QCI, the UE is also more severely impacted by the botnet than are the other legitimate devices that start transmitting before the attack. With a single attacker sending low QCI traffic, the legitimate device's throughput is cut to 48% of its requested rate, whereas the other legitimate UEs are able to achieve 62.5% of their desired throughput. A botnet of 15 attackers reduces the UE's throughput to 0.1%, at 190 b/s. Although the impact of the botnet is worse when the attackers send low QCI traffic while legitimate devices send high QCI traffic, everyone sending high QCI traffic generates significantly adverse consequences for the single legitimate device as well. When in the same cell as a botnet of 20 devices, the single legitimate user is granted only 34% of the bit rate it can send when there is a botnet of one device. This single client is therefore penalized more than the bulk of legitimate devices in the same cell that are allocated about 49% of their bit rate during an attack with one malicious device. Additionally, when the size of the botnet reaches 10 attackers, the single legitimate device's throughput reaches that of the other legitimate clients, even though it is requesting twice the bit rate. These results indicate that the network views the botnet as legitimate and therefore doesn't reduce the resources allocated to the botnet when the legitimate user attempts to transmit during the attack.

In our third set of simulation experiments, we examined the impact of a botnet on a densely-populated cell made up of 200 legitimate devices sending 20,000 b/s of uplink traffic and a varying number of attackers transmitting 2Mbps each. Clearly the results in Table 3 (a) and (b) indicate that a much smaller botnet can cause even more drastic ramifications than in a lightly used cell. This phenomenon is due to the eNB trying to accommodate the malicious UEs by reducing the throughput of devices already present in the cell. For example, it takes only one attacker sending low QCI traffic to spark a complete DoS for 98% of the legitimate population, with the remaining 2% of devices able to throughput only 0.5% of their requested bit rate.

Table 3 Uplink throughput for legitimate UEs in a densely-populated cell for different QoS traffic and varying botnet size, (a) Legitimate traffic: QCI 2, (b) Legitimate traffic: QCI 9

Size of Botnet [number of devices]	Attacker Traffic: QCI 9	Attacker Traffic: QCI 2
	[percentage of legitimate devices: uplink throughput]	
1	2%: 105 b/s	2%: 104 b/s
	98%: 0 b/s	98%: 0 b/s
2	2%: 105 b/s	2%: 104 b/s
	98%: 0 b/s	98%: 0 b/s
5	2%: 105 b/s	2%: 104 b/s
	98%: 0 b/s	98%: 0 b/s
10	2%: 105 b/s	2%: 104 b/s
	98%: 0 b/s	98%: 0 b/s
15	2%: 104 b/s	2%: 104 b/s
	98%: 0 b/s	98%: 0 b/s
20	2%: 100 b/s	2%: 100 b/s
	98%: 0 b/s	98%: 0 b/s

(a)

Size of Botnet [number of devices]	Attacker Traffic: QCI 9	Attacker Traffic: QCI 2
	[percentage of legitimate devices: uplink throughput]	
1	100%: 3,800 b/s	100%: 3,100 b/s
2	100%: 3,800 b/s	100%: 2,400 b/s
5	100%: 3,750 b/s	100%: 821 b/s
10	100%: 3,660 b/s	100%: 268 b/s
15	100%: 3,567 b/s	66%: 103 b/s
		34%: 0 b/s
20	100%: 3,500 b/s	57%: 103 b/s
		43%: 0 b/s

(b)

Similarly, when legitimate devices send high QCI traffic and attackers low QCI traffic (Table 3 (b)), it takes only one attacker to reduce legitimate throughput by 85%, yet in a lightly used cell, the same reduction requires 15 attackers. It is interesting to observe that when legitimate clients send low QCI traffic, the impact of the botnet is the same regardless of the QoS requirements of the attack traffic, as shown in Table 3 (a). However, when legitimate users send high QCI traffic, a botnet sending low QCI traffic causes a significantly worse impact as its scale increases than one sending high QCI traffic. In both cases, nevertheless, it takes only one attacker to reduce the legitimate throughput by 81–84%.

Figure 3 gives some insight on how legitimate user QoS is impacted by a botnet as it scales up, measured in terms of uplink packet latency between the device and eNodeB. In this experiment, legitimate users in both lightlyused (20 UEs) and densely-populated (200 UEs) cells attempt to send QCI 9 uplink traffic while a botnet of varying size sends QCI 2 traffic. As the results indicate, the larger the botnet, the greater the impact. It is interesting to note that it takes a smaller botnet to adversely affect legitimate users in a denselypopulated cell than it does in a lightly-used cell. Additionally, a botnet with only one device can produce such a high packet latency for legitimate clients sending QCI 9 traffic that it would result in their applications, such as web browsing or FTP, appearing unresponsive or even being unusable.

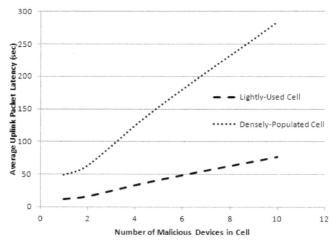

Figure 3   Average uplink packet latency over the air interface during an attack of variable sized botnets on densely-populated and lightly-used cells

5 Mobile Application Implementation and Results

This section describes the implementation of an Android application and a set of tools we have built to assess the effects of such attacks on existing infrastructure, along with an analysis of our experimental results. We used multiple smartphones connected to the same 4G LTE antenna and observed the transmission rate of a given file sent from a legitimate device to a server as we modified the traffic generated by the other surrounding devices. For every scenario, we recorded the data transmitted by the legitimate phone and used Speedtest.net App [5] to perform allocated bandwidth tests.

We implemented and deployed an application on a legitimate device that uploads a file to a server through a TCP connection and used a network capture tool (*tcpdump*) on the device to record the rate of traffic leaving the phone. Two separate devices acted as network flooders that utilized a UDP network traffic generator [2] and were configured to send traffic to a destination different than that of the smartphone. We captured the upload bit rate originating from the legitimate device and recorded the elapsed transmission time on each attempt. We also ran bandwidth tests for different network load on the antenna.

Table 4 shows the legitimate device file transfer rate we obtained during an attack with one and two malicious flooding devices as compared to during normal operations. We observe a degradation of the service between the non-flooded antenna test and the flooded antenna tests both in terms of bandwidth and response time. We observe an increase up to 19ms of response time (e.g., Ping test) and a diminution of upload bandwidth up to 3.64Mbps.

Next, we ran our flooder during an ongoing legitimate file transfer to observe the impact of a suddenly loaded antenna. Figure 4 shows the upload throughput for a file transfer on the legitimate device under normal circumstances (from 0s to 53s), while the antenna is flooded by one device

Table 4 Legitimate device file transfer rate results with one and two malicious flooding devices compared to during normal usage

	Average upload rate for a file transfer	Speedtest.net App Results
Normal Usage	5.12Mbps	Upload: 6.43Mbps Ping: 74ms
One Flooder	3.39Mbps	Upload: 3.63Mbps Ping: 93ms
Two Flooders	2.32Mbps	Upload: 2.79Mbps Ping: 84ms

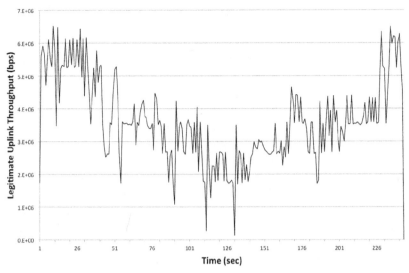

Figure 4   Legitimate throughput for a file upload through an antenna with different load over time

(from 53s to 109s), two devices (from 109s to 161s), again with one device (from 161s to 228s), and again under normal load (from 228s to 244s). We can observe a drop in the upload throughput while the flooders are transmitting, followed by a return to normal when they are stopped. In the worst case, we see normal throughput reduced up to 12,000 times when there are 2 flooders transmitting.

Although our results using existing infrastructure are small-scale, they prove the feasibility of our proposed attack strategies and show that a single malicious device has the potential to degrade legitimate user operations.

6 Conclusion

The rising popularity of smartphones has proliferated an abundance of mobile malware that could potentially perform large-scale coordinated attacks on communication infrastructure. Due to the limited frequency spectrum available to cellular networks, the physical layer is potentially vulnerable to denial of service attacks that can impact the bandwidth availability to all users in a cell. In this paper we examine several DoS attack strategies against the LTE RAN and study how the MAC uplink scheduler enables certain flavors of the attack, depending on the requested QoS of clients and the population of a cell, to be more effective than others. We study a variety of attack strategies in which we vary the traffic QoS requirements of legitimate and malicious devices in lightly-used and densely-populated cells for increasing botnet sizes. Our simulation results indicate that a single attacker is capable of significantly reducing the QoS experienced by legitimate devices in the same cell and, using certain strategies, inducing a complete denial of service for those clients. Since the network views the malicious UEs as benign, it tries to accommodate them as much as possible by assigning resources and reducing the throughput of legitimate devices present in the cell. We follow our sim-ulation experiments with a real working prototype on the Android platform that proves the feasibility of our proposed attack strategies and demonstrates the impact legitimate users can experience during the attacks.

References

[1] Gartner Says the Internet of Things Installed Base Will Grow to 26 Billion Units By 2020. http://www.gartner.com/newsroom/id/2636073.
[2] gen-send: A Simple UDP Traffic Generater Application. http://www.citi. umich.edu/projects/qbone/generator.html.
[3] Global LTE Subscription Growth. http://www.4gamericas.org/index.cfm ?useaction=page&pageid=2197.
[4] OPNET Modeler. http://www.opnet.com/solutions/network_rd/modeler. html.
[5] Speedtest.net App. http://www.speedtest.net/mobile/.
[6] LTE eNodeB MAC Scheduler Interface. White paper, Roke, 2009. http://www.roke.co.uk/resources/datasheets/108-lte-mac-scheduler-inter face.pdf.
[7] 3rd Generation Partnership Project; LTE; Technical Specification Group Services and System Aspects. Policy and charging control architecture; 3gpp ts 23.203. v12.3.0, 2012.
[8] Adam J Aviv, Katherine Gibson, Evan Mossop, Matt Blaze, and Jonathan M Smith. Smudge attacks on smartphone touch screens. In *Proceedings of the 4th USENIX conference on Offensive technologies*, pages 1–7. USENIX Association, 2010.
[9] R. Bassil, I.H. Elhajj, A. Chehab, and A. Kayssi. A resource reservation attack against lte networks. In *Communications and Information Technology (ICCIT), 2013 Third International Conference on*, pages 262–268, June 2013.

[10] Enrico Cambiaso, Gianluca Papaleo, Giovanni Chiola, and Maurizio Aiello. Slow dos attacks: definition and categorisation. *International Journal of Trust Management in Computing and Communications*, 1(3): 300–319, 2013.

[11] David Dagon, Manos Antonakakis, Kevin Day, Xiapu Luo, Christopher P Lee, and Wenke Lee. Recursive dns architectures and vulnerability implications. In *NDSS*, 2009.

[12] E. Dondyk and C.C. Zou. Denial of convenience attack to smartphones using a fake wi-fi access point. In *Consumer Communications and Networking Conference (CCNC), 2013 IEEE*, pages 164–170, 2013.

[13] William Enck, Patrick Traynor, Patrick McDaniel, and Thomas La Porta. Exploiting open functionality in sms-capable cellular networks. In *Proceedings of the 12th ACM Conference on Computer and Communications Security*, CCS '05, pages 393–404, New York, NY, USA, 2005. ACM.

[14] Adrienne Porter Felt, Matthew Finifter, Erika Chin, Steve Hanna, and David Wagner. A survey of mobile malware in the wild. In *Proceedings of the 1st ACM Workshop on Security and Privacy in Smartphones and Mobile Devices*, SPSM '11, pages 3–14, New York, NY, USA, 2011. ACM.

[15] Anthony Freed. Misosms malware sends your text messages to attackers in china, 2013. http://www.tripwire.com/state-of-security/top-security-stories/misosms-malware-sends-text-messages-china/.

[16] Nicola Gobbo, Alessio Merlo, and Mauro Migliardi. A denial of service attack to gsm networks via attach procedure. In *Security Engineering and Intelligence Informatics*, pages 361–376. Springer, 2013.

[17] George Hulme. Geinimi android malware has 'botnet-like' capabilities, 2011. http://www.csoonline.com/article/650866/geinimi-android-malware-has-botnet-like-capabilities?source=rss_cso_exclude_net_net.

[18] Woongryul Jeon, Jeeyeon Kim, Youngsook Lee, and Dongho Won. A practical analysis of smartphone security. In *Human Interface and the Management of Information. Interacting with Information*, pages 311–320. Springer, 2011.

[19] M. Khosroshahy, Dongyu Qiu, and M.K. Mehmet Ali. Botnets in 4g cellular networks: Platforms to launch ddos attacks against the air interface. In *Mobile and Wireless Networking (MoWNeT), 2013 International Conference on Selected Topics in*, pages 30–35, 2013.

[20] Kameswari Kotapati, Peng Liu, and Thomas F LaPorta. Cata practical graph & sdl based toolkit for vulnerability assessment of 3g networks.

In *Security and Privacy in Dynamic Environments*, pages 158–170. Springer, 2006.

[21] Mohit Lad, Ricardo Oliveira, Beichuan Zhang, and Lixia Zhang. Understanding resiliency of internet topology against prefix hijack attacks. In *Dependable Systems and Networks, 2007. DSN'07. 37th Annual IEEE/IFIP International Conference on*, pages 368–377. IEEE, 2007.

[22] Lam, Wayne. Wireless Communication Report-4G-LTE Landscape. https://technology.ihs.com/413870/wireless-communications-report-4g-lte-landscape-2012.

[23] LTE; Evolved Universal Terrestrial Radio Access (E-UTRA). Medium access control (mac) protocol specification. 3gpp ts 36.321. v12.0, 2013.

[24] Claudio Marforio, Aurélien Francillon, Srdjan Capkun, Srdjan Capkun, and Srdjan Capkun. *Application collusion attack on the permission-based security model and its implications for modern smartphone systems*. Department of Computer Science, ETH Zurich, 2011.

[25] Collin Mulliner and Jean-Pierre Seifert. Rise of the iBots: 0wning a telco network. In *Proceedings of the 5th IEEE International Conference on Malicious and Unwanted Software (Malware)*, 2010.

[26] Collin Mulliner and Giovanni Vigna. Vulnerability analysis of mms user agents. In *Computer Security Applications Conference, 2006. ACSAC'06. 22nd Annual*, pages 77–88. IEEE, 2006.

[27] Alexios Mylonas, Stelios Dritsas, Bill Tsoumas, and Dimitris Gritzalis. Smartphone security evaluation-the malware attack case. *SECRYPT*, 11: 25–36, 2011.

[28] Phillip Porras, Hassen Sadi, and Vinod Yegneswaran. An analysis of the ikee.b iphone botnet. In AndreasU. Schmidt, Giovanni Russello, Antonio Lioy, NeeliR. Prasad, and Shiguo Lian, editors, *Security and Privacy in Mobile Information and Communication Systems*, volume 47 of *Lecture Notes of the Institute for Computer Sciences, Social Informatics and Telecommunications Engineering*, pages 141–152. Springer Berlin Heidelberg, 2010.

[29] R. Racic, D. Ma, Hao Chen, and Xin Liu. Exploiting and defending opportunistic scheduling in cellular data networks. *Mobile Computing, IEEE Transactions on*, 9(5): 609–620, 2010.

[30] ComScore reports June 2012 U.S. mobile subscriber market share. http://www.comscore.com/Insights/Press_Releases/2012/8/comScore_Reports_June_2012_U.S._Mobile_Subscriber_Market_Share.

[31] Fabio Ricciato, Angelo Coluccia, and Alessandro DAlconzo. A review of dos attack models for 3g cellular networks from a system-design perspective. *Computer Communications*, 33(5): 551–558, 2010.

[32] S. Sesia, M. Baker, and I. Toufik. *LTE, The UMTS Long Term Evolution: From Theory to Practice*. Wiley, 2009.

[33] Muhammad Zubair Shafiq, Lusheng Ji, Alex X. Liu, Jeffrey Pang, and Jia Wang. A first look at cellular machine-to-machine traffic: Large scale measurement and characterization. In *Proceedings of the 12th ACM SIGMETRICS/PERFORMANCE Joint International Conference on Measurement and Modeling of Computer Systems*, SIGMETRICS '12, pages 65–76, New York, NY, USA, 2012. ACM.

[34] Patrick Traynor, Michael Lin, Machigar Ongtang, Vikhyath Rao, Trent Jaeger, Patrick McDaniel, and Thomas La Porta. On cellular botnets: measuring the impact of malicious devices on a cellular network core. In *Proceedings of the 16th ACM conference on Computer and communications security*, pages 223–234. ACM, 2009.

[35] Patrick Traynor, Patrick McDaniel, Thomas La Porta, et al. On attack causality in internet-connected cellular networks. In *Proceedings of 16th USENIX Security Symposium on USENIX Security Symposium*, pages 1–16. USENIX Association, 2007.

Biographies

Jill Jermyn is a PhD student at Columbia University, where she works under Professor Salvatore Stolfo in the Intrusion Detection Systems Lab. Some of her research interests are wireless and cellular network security, mobile, and cloud computing. Part of her previous experience includes internships at AT&T Security Research Center and IBM Watson Research Center. Starting

Fall 2014 she will be Adjunct Professor of Computer Science at Purchase College. Jill has been granted numerous awards for her work, including several from Google, Facebook, Applied Computer Security Associates (ACSA), Brookhaven National Laboratory, and the National Physical Science Consortium. Prior to her career in computing, Jill pursued a career as a concert violinist. She has performed at venues such as Carnegie Hall, Lincoln Center, Kennedy Center, the Austrian Cultural Forum NY, and the Los Angeles County Museum of Art, to name a few.

Gabriel Salles-Loustau is a PhD candidate in the 4N6 Cyber Security and Forensics Laboratory in the Electrical and Computer Engineering Department at the University of Miami. His research interests include systems and network security, mobile devices systems security and data privacy.

Saman Zonouz is an Assistant Professor in the Electrical and Computer Engineering Department at the University of Miami (UM) since August 2011, and the Director of the 4N6 Cyber Security and Forensics Laboratory.

He has been awarded the Faculty Fellowship Award by AFOSR in 2013, the Best Student Paper Award at IEEE SmartGridComm 2013, the EARLY CAREER Research award from the University of Miami in 2012 as well as the UM Provost Research award in 2011. The 4N6 research group consists of 1 post-doctoral associate and 8 Ph.D. students, and their research has been funded by grants from NSF, ONR, DOE/ARPA-E, and Fortinet Corporation. Saman's current research focuses on systems and smartphone security and privacy, trustworthy cyber-physical critical infrastructures, binary and malware analysis, as well as adaptive intrusion tolerance architectures. Saman has served as the chair, program committee member, and a reviewer for international conferences and journals. He obtained his Ph.D. in Computer Science, specifically, intrusion tolerance architectures, from the University of Illinois at Urbana-Champaign in 2011.

Triton: A Carrier-based Approach for Detecting and Mitigating Mobile Malware

Arati Baliga[1], Jeffrey Bickford[2]
and Neil Daswani[3]

[1] *NYU Polytechnic School of Engineering*
[2] *AT&T Security Research Center*
[3] *Twitter Inc.*

Received 28 February 2014; Accepted 28 February 2014;
Publication 2 July 2014

Abstract

The ubiquity of mobile devices and their evolution as computing platforms has made them lucrative targets for malware. Malware, such as spyware, trojans, rootkits and botnets that have traditionally plagued PCs are now increasingly targeting mobile devices and are also referred to as mobile malware. Cybercriminal attacks have used mobile malware trojans to steal and transmit users' personal information, including financial credentials, to bot master servers as well as abuse the capabilities of the device (e.g., send premium SMS messages) to generate fraudulent revenue streams.

In this paper, we describe Triton, a new, network-based architecture, and a prototype implementation of it, for detecting and mitigating mobile malware. Our implementation of Triton for both Android and Linux environments was built in our 3G UMTS lab network, and was found to efficiently detect and neutralize mobile malware when tested using real malware samples from the wild. Triton employs a defense-in-depth approach and features: 1) in-the- network malware detectors to identify and prevent the spread of malware and 2) a server-side mitigation engine that sends threat profiles to an on-the-phone trusted software component to neutralize and perform fine-grained remediation of malware on mobile devices.

* This work was conducted while Arati Baliga was with the AT&T Security Research Center. Neil Daswani conducted this research while employed at Dasient. Twitter acquired Dasient in January 2012.

Journal of Cyber Security, Vol. 3 No. 2, 181–212.
doi: 10.13052/jcsm2245-1439.324
© 2014 *River Publishers. All rights reserved.*

1 Introduction

Mobile devices have become an integral part of our daily lives; we rely on them to send and receive email, communicate with family and friends, perform financial transactions, and much more. Due to the inherent trust users place in these devices, as well as the availability and frequent download of hundreds of thousands of apps, it is no coincidence that mobile devices are now targets of complex malware attacks. According to a threat report by F-Secure Labs, 5,033 malicious Android applications were discovered in the second quarter of 2012, a 64% increase compared to the previous quarter, including the first Android malware to use a drive-by-download vector for infection [6].

From a mobility network provider point of view, the mobile malware threat not only impacts its individual customers, but also impacts the security and reliability of the mobility network as a whole. Researchers have shown the feasibility of denying mobility network services using specially targeted SMS messages, control channel vulnerabilities, and mobile botnets [28, 42, 51]. In fact, in the recent outbreak of SpamSoldier [14], attackers formed a botnet of SMS spammers, making what was once a research problem now a reality. As with many other security problems, the problem of mobile malware needs to be addressed holistically via a defense-in-depth approach that includes prevention, detection, containment, and recovery techniques. Various solutions have been proposed to tackle the increasing number of mobile threats, though from the perspective of a network provider, no optimal solution exists today.

Many app stores are beginning to use security APIs to scan apps before they list them [9] [1]. While this will limit some malicious apps from being downloaded by the user, mobile malware can be delivered via other attack vectors, such as visiting infected websites (drive-by-downloads), downloading apps from unsafe app stores, spam email or SMS/MMS, or simply downloading unsafe content from unrated or malicious web sites. Recent work has also shown the feasibility of subverting the app store review process, thereby compromising the integrity of the app store itself [47, 8]. Though most companies provide mobile variants of their signature-based anti-virus software; these schemes typically require an exhaustive set of signatures and can be easily thwarted by malware that use techniques such as encryption and packing [24, 34]. In fact, Google's own App Verification Service, introduced in Android 4.2, only detects 15% of 1,200 malware samples previously released to the public [36]. Alternatively, host-based behavioral detection engines, which can detect these sophisticated threats, are simply infeasible to deploy on current mobile devices due to their heavy resource requirements and limited energy constraints [46, 20].

To protect both their customers and network infrastructure, network providers frequently deploy network-based anomaly detectors capable of detecting malicious traffic patterns, such as botnet communication patterns, worm traffic, and DDoS attacks [40]. Within the mobility network, these same security services exist, though they are expanded to include mobility specific attacks, such as SMS spamming campaigns and premium number fraud. Traffic characteristics and malicious payloads are typically analyzed using inhouse analysis environments and third-party cloud services without resource constraints. Traffic characteristics from millions of users are analyzed every day, giving a network provider visibility into a large set of attacks. However, when the network detects a misbehaving device and determines that its activity is harmful to both other customers and the network, the only current possible mitigation strategy is deactivating the device, resulting in dissatisfied customers and calls to customer service.

In this paper, we describe Triton, a new, network-based architecture and a prototype implementation of it for detecting and mitigating mobile malware. Triton combines the strength of network-based detection with the abilities of a trusted device component to identify the malicious app and mitigate the infection. Triton employs a defense-in-depth approach where in-the-network malware detectors communicate network threats to an on-the-phone trusted software component to identify and neutralize malware on the device. Combining network based detection with the ability to identify malware on the device allows Triton to provide protection against threats even in the absence of an anti-virus signature, provide faster response to ongoing threats, operate at lower costs, and leads to a minimal increase in battery consumption on the end device.

The contributions of this work are as follows:

- The Triton architecture detects and renders malware ineffective. Triton derives its effectiveness by placing network level components that detect and communicate threat profiles to a trusted software component running on the device that can identify and mitigate malware.
- A prototype implementation of Triton that we built in our 3G UMTS lab, and discuss some of the real-world trade-offs that we encountered in building it.

The remainder of the paper is organized as follows. In Section **1**, we provide a quick primer of the 3G mobility network to provide the reader an understanding of how Triton fits in. In Section 3, we describe our defense-in-depth approach including Triton's design and implementation. We present our experimental results in Section 4. We address counter attacks, scalability and limitations in Section 5. Related work is covered in Section 6 and we finally conclude in Section 7.

2 Background

This section describes the basic elements involved in the 3G UMTS network [16]. We use this type of network to design, develop and test our architecture. The architecture that we propose is generic enough and can be deployed with other types of 3G networks as well as 4G LTE networks.

2.1 3G UMTS Network Primer

Figure 1 shows the basic components of a UMTS network. In a UMTS network, a mobile device connects to the network via a radio link to the nearest base station, also referred to as the Node B. Multiple base stations are connected to a Radio Network Controller (RNC). For access to the circuit switched services, such as phone calls and SMS messages, multiple RNCs are connected to a Mobile Switching Center (MSC). SMS messages are sent to the nearest SMS Center (SMSC) from the MSC over the control channel. For access to the data services, multiple RNCs are connected to the Serving GPRS Support Node (SGSN). The MSC, SGSN and the Visitor Location Register (VLR) track devices that are connected to the network that they are visiting. Every subscriber in the UMTS network is identified with an International Mobile Subscriber Identity (IMSI) number. Every device is identified with an

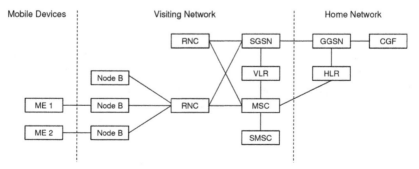

Figure 1   3G UMTS Network Architecture

International Mobile Equipment Identity (IMEI) number. Every subscriber also has a home network that stores the subscriber profile in the Home Location Register (HLR), including the IMSI and the IMEI numbers. Mutual authentication between a mobile device and a visited network is carried out with the support of the serving SGSN or the MSC/VLR. The Gateway GPRS Support Node (GGSN) acts as an anchor for all data traffic originating from the mobile device irrespective of its location. For simplicity and clarity, we have only described elements that are sufficient for basic understanding of the UMTS network. The GGSN is the component that our architecture interfaces with within the UMTS network and therefore is described in more detail below.

2.2 Gateway GPRS Support Node (GGSN)

The GGSN is a node that acts as a gateway between the mobility network and the Internet. The mobile device connects to the local SGSN, which in turn builds a tunnel to the GGSN using the GPRS Tunneling Protocol (GTP). When the device moves to a different location, it switches SGSNs while the GGSN serves as the anchor point of the tunnel that routes data traffic to the Internet. The GGSN covers a very large part of the mobility network for data services as it resides within the user's home network. This placement allows the GGSN to serve as a central point of observation for data traffic and therefore is ideal for placement of network based malware detectors.

The GGSN assigns an IP address to every single outgoing data connection originating from the mobile device. This IP address is randomly picked from a pool of IP addresses owned by the GGSN. Therefore, a single IP address from this pool might represent different mobile devices within the same mobility network at different points in time. Alternatively, different IP addresses might correspond to the same device at different time instances. The GGSN contains all information about user's data usage. It feeds this information into a Charging Gateway Function (CGF). The CGF accounts for data usage and generates billing information based on the usage and the type of data plan subscription. The GGSN naturally acts like a Network Address Translation (NAT) device and has information to correlate the IP address assigned to the device to its identity at any given point in time. This structure of the mobility network provides the following advantages that we leverage.

- **Centralized view of data traffic:** The GGSN acts as a gateway to the data traffic and has visibility into data generated by all the mobile devices that it serves. This centralized view enables detection of large scale attacks, such as command and control traffic originating from mobile botnets and fast spreading worms.

- **Mapping IP to the device identity:** When a new mobile device initiates a data connection, the GGSN creates a Packet Data Protocol (PDP) context for the device. The PDP context is maintained at both ends of the GTP tunnel between the SGSN and the GGSN. It contains all information about the device that has requested the data service including its IMSI/IMEI numbers. The GGSN picks a free IP address from its pool and assigns it to the new PDP context. Because the GGSN maintains this information, at any given point in time, it is able to identify the device with the given IP address. The architecture we propose utilizes this mapping to accurately identify infected devices sending out malicious traffic.

3 Design and Implementation

Triton employs a defense-in-depth approach in combating mobile malware, which broadly comprises of the following steps.

- **Infection prevention:** Triton is able to prevent infections by blocking mobile devices from either visiting or downloading known "bad" apps and content. It can push infection information of new and ongoing threats to non-infected devices, which will prevent malware from running on these devices altogether.
- **Effective detection:** Mobile devices are heavily network centric as most content is delivered from the Internet via apps or websites. Triton places malware detectors within the mobility network and monitors for signs of infection. It also interfaces with third-party cloud services for specialized offline analysis of apps and content, which it uses for detection and prevention.
- **Immediate containment:** Triton employs a trusted component on the mobile device that receives threat profiles from the network. Threat profiles characterize the traffic that was flagged by the network as malicious. The trusted component can identify the application that generated the malicious traffic and immediately stop it from executing. It can also notify the user and remove the application from the device once it has been neutralized.
- **Fine grained response:** Triton's fine-grained response of only containing malware without hampering execution of other legitimate applications, allows the user safe continued access to his device.

3.1 Architecture

Figure 2 shows the end-to-end architecture of Triton and how it interfaces with the GGSN within the 3G UMTS network. Triton places several new components within the mobility network - the Mitigation Engine (MiE), the Network Malware Detector (NMD), the Filter, the Packet Inspector (PI) and the database. It also places a Trusted Host Component (THC) on the mobile device itself to assist with containment.

As explained in Section 2.2, mobile devices send data traffic to the Internet via the GGSN. When a mobile device attaches to the network, it establishes a new PDP context and is assigned an IP address from a pool owned by the GGSN. We instrument the GGSN to log the IP address assigned to the device and its IMSI/IMEI numbers into the database. It also logs the time duration for which the current IP belonged to the specific device, i.e. the duration of the PDP context. This information is required to map the IP address assigned to the device with its device identity (IMSI/IMEI).

The PI sniffs the Internet facing interface of the GGSN and logs all network data flows, originating from and to the mobile devices, into the database. It also logs information from some application level protocols, such as HTTP and DNS. The NMD operates on logged network flows and detects malicious traffic patterns e.g., the NMD might find a mobile device conducting an IP scan or a port scan, or sending traffic to blacklisted command and control servers. Since the NMD has a large scale view of the mobility network, it is well placed to detect spread of worms or other types of large scale attacks, such as DDoS or botnet command and control communication patterns. It is also well-placed to detect other kinds of application level accesses, such as access to malicious websites or malicious app downloads. Along with traditional network-based detectors, the NMD can also rely on external cloud-based

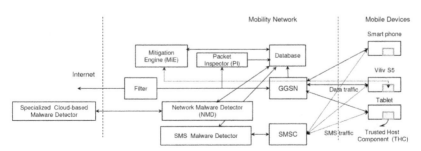

Figure 2   Mobile malware mitigation architecture

services for specialized analysis, e.g., analysis of specific URLs for drive-by-downloads or conducting behavioral analysis of an app that was accessed by the end mobile device. Analysis results from cloud services are used as network signatures for detection both in the NMD and the Filter. The Filter component blocks future access from mobile devices to websites or apps that are identified as malicious. The MiE uses the alert information generated by the NMD to generate a threat profile and communicates the threat profile to the THC on the end mobile device. The mobile device identifies the application that is infected and immediately stops it from executing on the device.

Below, we explain in detail the role of each of the components of Triton and how they communicate with each other to effectively combat mobile malware.

3.1.1 Trusted Host Component (THC)

Triton places the THC on the mobile device in two different ways; primarily based on whether the underlying platform supports virtualization.

Figure 3(a) shows a THC on the device that supports virtualization. On a virtualized platform, the THC runs as an application inside a privileged virtual machine (VM) and all other user applications run inside a User VM. Placing the THC in this fashion allows us to leverage certain well-known security properties of the virtual machine architectures [29, 27]. By placing the THC in the privileged VM, it is effectively isolated from other applications running on the device yet is able to inspect on the state of the applications and the operating system running inside the User VM. This architecture can protect against malware that resides in user, as well as kernel, space inside the User VM.

Figure 3(b) shows the THC running inside the operating system kernel for platforms that do not support virtualization, which is the case with current

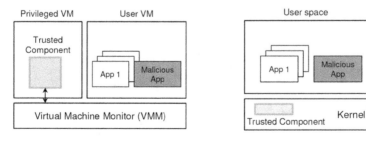

(a) Trusted component in the privileged VM   (b) Trusted component inside the operating system.

Figure 3   Mobile device architecture and placement of the trusted component

commercial mobile phones. We use this architecture on smart phones where the THC runs inside the kernel. By running in kernel space, the THC is able to intercept system calls from user space applications and inspect application state. However, this model has the limitation that it is unable to detect or defend against attacks that compromise the kernel, e.g., kernel-level rootkits. While using this architecture, we assume that the threat to the device is only in user space.

In both cases, the THC listens for connections from the MiE on a special TCP port. It also maintains a rolling log of network connections both incoming and outgoing generated by user applications running on the device. This log maps each network flow to the application that generated the flow or is the recipient of the flow. When the network identifies that a device is infected with malware, it receives infection information from the MiE about the network flows that were found to be malicious. The THC refers to its log to find the application that generated the malicious flow.

3.1.2 Packet Inspector (PI)

The PI component sniffs the outgoing physical interface of the GGSN, also known as the Gi interface. It logs into the database, all bidirectional network flows from the mobile devices. Network flows comprise of source and destination IP addresses, source and destination ports and the protocol used. The PI also logs some relevant information from certain application level protocols of interest, such as domain names for DNS and URLs for HTTP. DNS information allows the NMD to identify applications trying to contact malicious domains. Logging HTTP URLs helps identify access to malicious websites from mobile devices.

3.1.3 Network Malware Detector (NMD)

The NMD operates on logged network flows and identifies malicious network traffic. The NMD is designed with extensibility in mind and can leverage any number of well known tools or techniques to identify malicious traffic patterns [2, 15, 23, 31–33, 41 52–53]. In order to improve Triton's effectiveness over time, future network-based detection algorithms can be easily incorporated into the NMD. At the IP/TCP/UDP layer, the detector might find a device scanning other IP addresses or ports. A high scan rate indicates a worm trying to spread to new devices, such as the iKee.B worm which targeted jailbroken iPhones [49]. Often port scans look for specific open ports running services with vulnerabilities. Other types of network layer detection might involve using the amount of data or connections to detect Denial of Service (DoS) attempts, including DoS attempts on the mobility network itself [42]. The NMD can also maintain an IP blacklist, obtained from 3rd party sources, network based detection algorithms and internal observations, to identify malicious connections from devices to such servers. This often indicates infection, e.g., a bot trying to connect to its command and control server or a malicious application that tries to send stolen information back to the attack server.

Apart from detection at the network layer, detection can be incorporated by scanning headers of higher level protocols contained within the network traffic, such as DNS. With DNS information, the network detector can identify malicious connections to blacklisted domains. In addition to scanning higher level protocol headers, the detector can also perform deep packet inspection on suspicious traffic to match for known malware signatures. This approach can be used selectively as it increases the storage load on the server significantly.

The NMD logs all alerts into the database. The alert contains the IP address of the device and the time at which the flow was recorded to be malicious.

Specialized detectors. Network providers typically rely on specialized detectors for certain services that are outside their area of expertise. In Triton, the NMD employs specialized cloud-based detectors for a more comprehensive, time-consuming analysis of unknown URLs and apps seen in the network. When a user downloads an unknown app or visits an unknown URL, the Filter optimistically grants access to the resource, while initiating a URL/app scanning request to the cloud service in parallel. Though some simple, signature-based checks can be done to determine if an application or URL is already previously known to be malicious, scanning applications and URLs for previously unknown threats using dynamic or behavioral analysis can detect malware variants for which no signature exists yet. Analysis results returned by the cloud service are fed back into the NMD to enhance its algorithms and to prevent future accesses to the malicious URLs/apps. If a device gets infected before the results are received, an alert is generated by the NMD, and the mitigation steps to contain the effects of the malware are executed by the MiE and THC working together in tandem.

SMS Malware Detector. If the cloud-based analysis reports malware sending premium SMS messages or messages that have been previously identified as spam [3], the message or premium number is added to an SMS blacklist within our database. The SMS Malware Detector, running on the SMSC has access to this database and can match all outgoing SMS messages with this blacklist and generate an SMS alert.

3.1.4 Filter

The Filter component blocks access to malicious content/connections e.g., it can check for future accesses to URLs or apps that are flagged as malicious. In such a case, the Filter component drops the request and returns a stub page to the user, which informs him of the website being malicious. The Filter component can also filter generic netflows identified as malicious, such as communication with blacklisted IPs or domain names.

3.1.5 Mitigation Engine (MiE)

The MiE processes the alerts generated by the NMD. An alert contains the IP address of the infected device and the time at which the alert was generated for the given network flow. The MiE has to first identify the correct device based on this information as the device might have disconnected from the network or might have acquired a new IP address. The MiE first obtains the IMSI/IMEI of the device that the IP address belonged to at the time the alert was generated. This information can be obtained by correlating the time for which the PDP context was valid and had the aforementioned IP address. From the IMSI/IMEI number, it checks if the device is connected to the network and has a valid PDP context. It obtains the new IP address of the device in this case and makes a connection to the THC on the device. If the device is offline, this alert is ignored and processed later when the device connects back to the network.

With the valid IP address, the MiE connects to the THC on a special TCP port. The THC listens on this special port for commands from the MiE and receives threat profiles in order to identify malicious apps.

3.2 Network-Device Communication

When a device is uninfected, the network passively scans for signs of malicious traffic generated by the device. Communication is initiated by the MiE, only when signs of infection are found via the NMD. In all cases, the MiE initiates the connection to the THC on the device. We assume that the MiE and the THC have a unique preshared key that they use for secure communication. The preshared key can be securely stored on the device and distributed along with the THC.

3.2.1 Threat profile communication

The MiE identifies the infected device in response to an alert generated by the network detector or as a result of offline analysis. For example, the NMD might have identified that the device is infected because it found the device initiating an HTTP connection to a bot controller. After establishing a secure channel with the THC on the device, the MiE sends a threat profile to it. Figure 4 shows an example threat profile generated by the MiE for a device infected with the DroidDream malware [7]. In this case, the network identifies a device which has contacted a blacklisted IP of the botnet command and control server. The profile includes the details of the malicious flow as identified by the network, which includes the source and destination IP and port pairs, the protocol and the time the connection was made. This profile also commands the THC to terminate the application that generated the malicious flow.

The host component also maintains a log of network connections from the device. This log comprises of the eight tuple <Source IP, Dest IP, Source Port, Dest Port, Protocol, Application, Start Time, End Time>. Compared to the network, the host component additionally stores the application identity that is sending/receiving packets. By matching the tuples with the threat profile, the host component is able to find the application that generated the malicious flow.

The MiE responds with possible remediation actions to be performed depending on the severity of the threat. If the flow, identified by the network, is not found in the log, the THC stores this information in a watch list and looks for future matching packets that might be sent from the device. For example, if a bot on the device initiated HTTP connections to its controlling server, this event is likely to repeat within the near future and will be detected at that point in time.

```
<threat>
  <description>
      Blacklisted  IP  Connection
  </description>
  <profile>
    <flow>
      <sip >172.16.23.68 </sip>
      <sport >42965</sport>
      <dip >184.105.245.17 </dip>
      <dport >8080</dport>
      <protocol >TCP</protocol>
      <time >2011−08−01  15:52:35 </time>
    </flow>
  </profile>
  <mitigation  action="kill_app"/>
</threat>
```

Figure 4   Threat profile for the DroidDream app connecting the botnet server

3.2.2 Mitigation Actions

Mitigation actions can be explicitly requested by the MiE or performed by the host component after receiving infection information. In both cases, the mitigation actions effect only the malware program running on the device without hampering the user from using other applications or functionality. In some cases, where the network has enough information about an ongoing threat, it can request the THC to perform prevention in order to contain the threat. For example, if Triton has identified both the network activity and application of a fast spreading worm, it might preemptively share this threat profile with other devices before they are infected, thus preventing infection.

Android smart phones today allow Google to remotely send commands to either install or remove applications from their smart phones via their GTalk service. Google has been using this feature to revoke applications found to be malicious or violating their Android market developer distribution agreement or content policy [13, 5]. Our architecture on the other hand allows the mobility carrier to invoke such functionality on all types of heterogeneous devices that are allowed to connect to its mobility network, upon discovering that those devices are infected. As opposed to the kill switch that Google can invoke to remove applications only downloaded from the Android market [5], Triton can protect end users from malicious apps that might have been installed from other alternative app stores, malware that gets installed as a result of a drive by download, malware that installs with the user permission as a result of email or SMS/MMS spam or other types of malicious attacks, such as worms, etc.

3.3 Implementation

We have a fully operational 3G UMTS instance that implements the Triton architecture. Below, we describe the implementation details of prototyping Triton within our 3G UMTS lab network.

3.3.1 Network components

We use the OpenGGSN software running on a Linux server as the GGSN node [12]. The OpenGGSN node interfaces with the SGSN within our 3G wireless lab. We modified OpenGGSN to log information about PDP contexts assigned to the mobile devices within the mobility network into a MySQL database that runs on the same server. This information consists of the device identity, i.e. the IMSI/IMEI numbers and its corresponding IP address for a valid PDP context time duration. The PI, NMD, MiE and the Filter are all applications that run on the same Linux server and can query the MySQL database.

3.3.2 Trusted component on Android

On Android, the THC runs within the Android kernel, implemented as a kernel module as shown in Figure 3(b). We use this architecture to demonstrate the implementation on smart phones as they exist today. Because the THC runs within the kernel, it can only provide complete protection against attacks that operate in user space, which is largely the case for smart phone malware today. This implementation does not take into account attacks that already have obtained kernel level control, such as kernel-level rootkits. We defer discussion about how malware might subvert the host component if it obtains kernel level control and our countermeasures against such attacks to Section 5.1.

The THC obtains information about Android applications by intercepting system calls. It intercepts all socket related calls to create a log of all TCP and UDP traffic that originates from Android applications. It buffers and logs network activity within a small rolling log. This log is comprised of the network end point information, such as source and destination IP addresses and port numbers, the application that generated or received the traffic and the start and end times of the network flow. The application name that sends traffic on a specific socket is obtained by internally walking the various kernel data structures. A significant amount of malware on Android also generates malicious SMS messages. To keep track of SMS messages, the THC also intercepts all writes to the modem file descriptor and looks for AT commands. By using this technique, it also logs all SMS messages that are generated by Android applications.

To receive threat profiles from the network MiE, the THC runs a TCP server within kernel space as a kernel thread. This thread listens to commands from the network component and can decode the threat profiles that it receives. Upon receiving a threat profile, it decodes the malicious flow, refers to its log to find the application that generated the flow and kills the application by posting a kill signal to the appropriate process. The application is further blacklisted and prevented from running on the device thereafter.

3.3.3 Trusted component on Xen/Linux

Current mobile virtualization solutions are limited in their availability and cannot be installed on any commercially available devices today. To demonstrate how Triton works with virtualizable platforms, which are expected to be available on smart phones in the near future [11, 17, 35], we built our prototype on the Viliv S5 mobile device due to its functional equivalence of a smart phone. The Viliv S5 is equipped with an Intel Atom Z520 1.33 GHz processor and runs the Xen VMM [27] to achieve isolation between the user VM and the THC executing within a privileged VM, both running the Linux operating system as shown in Figure 3(a).

The THC inside the privileged VM must identify network activity generated by the user VM and identify the application that generated this traffic. By using a hypervisor-based system, the THC can remain isolated and secure even if the user VM's kernel is compromised by malware. Because we assume the kernel in the user VM can be compromised, the THC does not rely on any information within the user VM kernel and therefore must bridge the semantic gap between the hypervisor and the user VM.

In the hypervisor, we add functionality for intercepting and forwarding process and networking related system calls to the THC in the privileged VM. We use a technique published in [19], to pass all system calls to the hypervisor. Relevant system calls are forwarded to the THC via shared memory pages, which are accessible to both the hypervisor and the THC. The THC waits on a Xen event channel in order to be notified when new information arrives on the shared pages. Below, we describe how this implementation tracks applications running on the system and how it accounts for the network traffic that they generate.

Process tracking. In Linux, processes are executed using the execve system call. The execve system call takes an argument, which is the full path name of the application to be executed. The THC obtains the name of the application executing by using this information from the hypervisor. To identify the process currently running, without relying on the kernel data structures of the user VM, we rely on the characteristics of the x86 memory management system.

In the x86 architecture, each process is separated within its own virtual address space using multi-level pages tables. Each address space is indexed by a high level page directory table, which is stored in the process control register (CR3) during process execution and can be used to uniquely identify a process [39, 18]. When a process is created using the the execve system call, we capture the address of the page directory and pass the address on to the THC. This gives us a mapping between a binary name and the current process running, identified by the CR3 value.

Network activity tracking. When an application generates network activity, it makes a sequence socketcall system calls. We trace various socket calls, such as connect and sendto, and pass their arguments to the THC in order to log network activity of the user VM. Since these system calls occur within the context of the currently executing process, the CR3 register is also passed to the THC in order to identify the application which caused this network activity. This generates a mapping between the application and the network activity that it is involved in.

Mitigation. Because killing a process relies on various kernel data structures of the user VM, the THC cannot directly kill a running application. Instead, once a malicious application has been identified, the binary name and any currently executing CR3 values corresponding to this binary are blacklisted. If the malicious application attempts to make any further socket system calls, the hypervisor returns an error resulting in a system call failure. This forces the kernel to return an error, effectively denying service to the application.

4 Evaluation

In this section, we describe the experimental setup and the specific experiments that we used to illustrate the defense-in-depth approach of our prototype Triton.

4.1 Experimental setup

Mobile devices equipped with specialized SIM cards can connect to the base station in our 3G wireless lab. The devices that we use in our experiments are the HTC Aria smart phone and the Viliv S5. The HTC Aria phone is equipped with a Qualcomm MSM 7227 Chipset, a ARMv6 600 MHz processor, and 291 MB of memory. We run the CyanogenMod 7 distribution [4] and Android version 2.3.4, API version 10 on the phone. The trusted component runs as a kernel module on this phone. The Viliv S5 is equipped with an Intel Atom Z520 1.33 GHz processor, 4.8" touch screen, 32GB hard drive, 1GB of memory, WiFi, Bluetooth, a 3G modem, and GPS. We use the Xen 4.0.1 hypervisor in paravirtualized mode. The privileged VM runs Fedora 12 with Linux 2.6.27.42 patched for Xen dom0 support while the User VM runs CentOS 5.5 with Linux 2.6.27.5 including Xen paravirtualization support.

4.2 Malware Prevention

We describe here the experimental setup that we have to prevent mobile devices from downloading known bad content and apps. Once resources are identified as malicious, future access is blocked, effectively protecting end users from downloading malicious content.

4.2.1 Behavioral Scanning of Apps and URLs

The Triton architecture uses a cloud-based malware analysis service to behaviorally scan applications and URLs offline. When a user downloads an application through an app store and/or has an application transmitted to his device, the download request URL, containing an application identifier, is observed by the packet inspector in the network. Triton uses this identifier to download the application through the normal app store channel and does not expose the user that has downloaded the application in any way. Due to privacy concerns, we do not rely on any sensitive information of the user and do not obtain applications from user devices. Triton currently only intercepts apps downloaded from the official Android market, but can be easily extended to include other app stores as well.

If an app has never been downloaded before or its status has expired, either the application identifier or application itself is sent to the cloud-based service for static and behavioral analysis. The static analysis phase involves quick checks against previously known malware signatures and identifies an application's possible characteristics based on permissions, API calls and app metadata. On the other hand, behavioral analysis of the application is conducted by emulating simulated user actions within a mobile emulator sandbox to determine what the application does. The analysis platform records various characteristics of an application such as system calls, bandwidth utilization and connections to domains/IPs. If the application sends SMS spam, attempts to gain root access, connects to a bot master, etc., the application is flagged as malware. Network characteristics, such as command and control servers or SMS spam messages, are logged in the database for use in the NMD.

In order to trigger malicious features of an app, user behavior is dynamically generated while the app is running. Depending on the complexity of the malware, multiple levels of scanning may be needed such as behavior after reboot, directed user-like behavior, and random aggressive crash testing. Though some apps may take up to 2 hours to analyze, in general we found that most malicious apps currently trigger their malicious functionality right after execution, in which case will be detected and generate a network signature within three minutes. The app analysis platform currently incurs a false positive rate of 1 in 10,000 and is continuously improving over time.

The Filter blocks an app download if the app was previously flagged as malicious. If the app has been downloaded before and was reported as benign, no new request is sent to the cloud service and the app can be downloaded as normal.

Similarly, when a user visits a URL on the mobile phone's browser, the URL is forwarded to the cloud service to scan for drive-by-downloads that might be initiated by that URL. The cloud service uses its malware analysis platform to conduct both static and behavioral checks to determine what the URL does, and the URL is flagged if it attempts to conduct a drive- by-download or is identified to be a phishing URL. The malware analysis platform used for URL detection had a low false positive rate, approximately 1 in 10,000,000. A similar caching mechanism with a set timeout is used to avoid redundant requests to the cloud service.

4.2.2 Scan Frequency and Caching

In order to put an upper bound on the number of users that can be infected in between scans of a resource, the frequency of scanning is chosen to be proportional to the frequency of accesses. Such an approach allows for an elegant trade-off between the scanning cost and level of security desired. If a resource is very popular, it is scanned more frequently as an infection of that resource could affect many users quickly, whereas if a resource is less popular, the cost of scanning it is kept in proportion with its low usage.

In our experiments, we send every new request that has not been refreshed within the past four hours to the cloud service for scanning. Upon the first visit of a malicious website or attempt to download a malicious app, we found that subsequent users are protected from visiting the website by the Filter component, which blocks access to the URLs found to be malicious.

4.3 Malware Detection and Mitigation

This set of experiments illustrates how Triton effectively detects and mitigates malware on the end mobile device. In all the experiments, we run the malware on the device and the network detector generates an alert as soon as it detects the first malicious flow. The MiE generates the threat profile and sends it to the THC on the device, which accurately identifies the malicious application and stops it from executing on the device. It is further blacklisted and prevented from running in the future.

4.3.1 Android malware

Table 1 shows the malicious apps tested, some of which were uploaded into the official Android market and were downloaded by users. We use these in our experiments as they form a good representative set of malware that exists today for Android phones. Column 1 shows the malware name. Column 2

Table 1 This table shows the malware that we use to run experiments on the HTC Aria smartphone running Android. All attacks were successfully detected by Triton and the malware was disabled from executing on the device.

Malware Name/Android Package Name	Malware Type	Malicious Network Behavior
Trojan: Android/Geinimi.A	Trojan, Bot	Connects to several malicious domains
Exploit:Linux/DroidRooter.A	Trojan, Bot	Connects to a malicious IP address
Trojan:Android/Twalktupi.A	Trojan, SMS spam	Connects to a malicious domain Sends spam SMS
Android.Trojan.Bgserv.A	Trojan, SMS spam	Sends SMS
Android.Trojan.FakePlayer.A	Trojan	Sends SMS to premium number
Trojan:Android/HippoSms.A	Trojan	Connects to malicious domains Sends SMS to premium numbers
Android.Trojan.GGTracker.A	Trojan	Sends SMS to premium number
Golddream.A	Trojan, Spyware	Connects to a malicious domain
Plankton	Spyware	Connects to a malicious domain

shows the type of malware that we obtained from our forensic analysis and documentation publicly available about the malware. Column 3 shows the network behavior of the application that was flagged as malicious.

All the malware samples were detectable in the network because they performed one or more of the following actions - (a) contacted malicious IP addresses (b) contacted malicious domains (c) Sent SMS spam or (d) Sent SMS to premium numbers.

The Trojan:Android/Geinimi.A is a repackaged sex positions application that connects to a botnet. The Exploit:Linux/DroidRooter.A, also popularly known as DroidDream, is a repackaged version of a bowling game application that also connects to a botnet and awaits additional commands.

The Trojan:Android/Twalktupi.A, popularly known as the "Walk&Text" application, sends SMS spam to all the contacts available in the contact list of the infected device. Android.Trojan.Bgserv.A also sends SMS spam. Android.Trojan.FakePlayer.A, Trojan:Android/Hippo-Sms.A and Android.Trojan.GGTracker.A all send SMS messages to premium numbers.

Plankton and GoldDream.A are spyware applications that connect to malicious domains and leak sensitive information, such as IMEI numbers. Gold- Dream.A logs incoming SMS messages to a local file and sends this over to the attacker. Plankton 1 was added to about ten other applications on the official Android market from three different developers. Its stealthy design also explains why some earlier variants have resided on the market for more than two months without being detected by current mobile anti-virus software [10]. Since Plankton[1] runs a background service that connects to a remote server, identified by a scanning request to the cloud service, all apps that include Plankton would be automatically detected by the NMD and disabled via the THC.

Triton successfully detected and mitigated all malware attacks on the end device.

4.3.2 Linux malware

On Linux, we chose malware that had some network footprint, such as worms, malware that launches Denial of Service (DoS) attacks, and bot software. Although the Linux platform is not a popular target for malware writers, we use real Linux samples and find that they do illustrate the capabilities of Triton.

Table 2 shows the representative samples that we used in our experiments. Column 1 shows the name of the malware. Column 2 shows the type and Column 3 shows the network behavior that the Malware Detector used to flag the flow as malicious and generate an alert. For IP and port scanners, the Malware Detector detects them based on the fact that the number of connections to a diverse set of IP addresses/ports exceeds a set threshold. Blacklisted IP addresses and ports are detected in a similar fashion as the Android malware, where the blacklists are constantly updated from third party cloud services doing malware analysis. DoS attempts are detected when malware tries to open connections to a given IP address or a set of IP addresses within a specified period of time that exceeds the normal threshold. In all the experiments with Linux malware from the set above, the malware detector successfully detected the malicious flows and the malware on the device was correctly identified from the threat profile generated by the MiE. The malware was subsequently disabled and blacklisted from running on the device in the future.

4.4 Performance

Below, we discuss the response time and the performance overhead of Triton.

---

[1]Prof. Jiang's research group at NC State Univ identified Plankton as spyware, but the authors of Plankton claim to be a legitimate ad network. Google suspended Plankton from its Android Market.

Table 2 This table shows the malware that we use to run experiments on the ViliV S5 running Linux inside a Xen VMM. The network behavior was flagged as malicious, and the malware was disabled from executing on the device.

Malware Name	Malware Type	Malicious Network Behavior
Trojan-DDoS.Linux.Fork	Bot	Connects to a blacklisted destination server and port number
Trojan-Spy.Linux.XKeyLogger.b	Port scanner, Keylogger	Conducts high rate TCP port scan
Net-Worm.Linux.Cheese	Worm	Connects to random IP addresses on a specific port
Net-Worm.Linux.Mworm.a	Worm	IP/Port scan over specific IP ranges
Trojan-DDoS.Linux.Blowfish	Worm	Connects to a blacklisted port
DoS.Linux.Arang	DoS	Creates a denial of service attacks to a specified victim
FTP AnoScan	Port Scanner	Scans for open FTP ports
FTPNullSearch02	Port Scanner	Scans for open FTP ports
Flooder.Linux.Alcohol.a	DoS	Creates a denial of service attacks to a specified victim

4.4.1 Response time

Automated mitigation has the potential to protect many users as well as a mobility provider's network from the negative effects of malware. For instance, in the DroidDream attack that took place in March 2011 in which over 260,000 users had downloaded malware to their phones, automated mitigation could have protected most users.

DroidDream was injected into re-packaged versions of over 50 of the most popular applications on Google's Android Market, and once downloaded, the malware shipped the user's IMSI/IMEI numbers and other personally identifiable information (PII) to a bot master. The attack was first noticed on March 1, 2011 [7], and Google started rolling out a fix in the form of the Android Market Security Tool on March 5, 2011. While data is not available regarding the download rate of DroidDream applications, by assuming a uniform download rate we can do a back-of-the-envelope calculation as to how many users could have been protected using automated mitigation.

As per the forensics for DroidDream, identified by Triton as shown in Figure 4, an infected phone can be identified by a connection to the DroidDream bot master, 184.105.245.17. Once DroidDream and its bot master are identified as malicious, Triton can contain malicious effects on all phones which download DroidDream thereafter. In particular, the mitigation engine would instruct the trusted component on the phone to kill the DroidDream process, and firewall off communication with the bot master. If 260,000 phones were infected by March 5, and the infections started on March 1, then approximately 2,200 phones were infected per hour. If a threat profile were to have been deployed by the mitigation engine within the first hour, then 99.2% of attempted DroidDream infections that occurred afterwards would not result in PII leakage or compromise of the phone. In addition, for the 0.8% of phones that were already infected, the DroidDream process could be killed and communication with the bot master could be cut off to prevent further damage.

4.4.2 Performance overhead

In this section, we run the following three workloads to record the overhead generated by running the THC on the device.

Browsing Workload. We use an automatic browsing script to measure our overhead against a typical mobile browsing experience. During the experiments we visited google.com, gmail.com with an account opened, and cnn.com. The script also watches a 60 second YouTube clip, within the browser on Linux and within the YouTube app on Android. Lastly the script checks an email account using the Thunderbird application on Linux and the default email application on Android.

To measure the overhead, we execute the workload on each platform both with and without the host component. We measure the time the workload executes using the wall clock time and average the results over five runs of the experiment. The results are summarized in Table 3. The in-kernel implementation on Android incurs a minimal overhead of 1.2%, while the VMM based implementation incurs a 9% overhead. The higher overhead is caused by the increased complexity of the hypervisor-based host component compared to the in-kernel version.

File Download. Mobile device users frequently download small files, such as music MP3s or applications from various app stores. To measure the overhead of a similar file download, we downloaded a 5 megabyte file from [21]. We performed this experiment on both the Linux and Android platforms by using the wget command. Because the file download initiates a single TCP connection we observe minimal overhead and report the download time overhead for both the kernel and hypervisor-based host component as 1.7% and 0.7% respectively. In each case the overhead is minimal and within the standard deviation of the workload.

CPU Intensive Workload. For completeness we include a CPU intensive workload for the hypervisor-based component. Because we trap every system call within the hypervisor, we wanted to observe the overhead of the system

Table 3 This table shows the performance overhead recorded for three different workloads resembling user actions on the device with and without the trusted host component (THC)

Workload	Android THC		Linux THC	
	Without THC	With THC	Without THC	With THC
Web Browsing	$150.2 + -1.6s$	$152 + -2s(1.2\%)$	$198.2 + -2.9s$	$216.4 + -8s(9.2\%)$
File download	$114 + -0.7s$	$114.2 + -1.6s(1.7\%)$	$114.4 + -3.2s$	$115.2 + -2.9s(0.7\%)$
LMBench	N/A	N/A	$198.3 + -0.6s$	$201.3 + -2s(1.5\%)$

in general. We chose a CPU intensive workload designed to measure OS performance called lmbench [44]. Lmbench exercises multiple OS interfaces and calls numerous system calls. We report an average overhead of (1.5%) and a standard deviation within both experiments, with and without the host component.

5 Discussion

In this section, we discuss counter attacks, scalability issues and the limitations of our approach.

5.1 Subverting the Host Component

The attacker might be able to subvert the host component in one of the following ways:

Root the device. On non-virtualized platforms, such as smart phones available today, the trusted host component runs as part of the operating system kernel. Several instances of malware on the Android platform have been known to root the device and install a rootkit [13], potentially compromising the integrity of our host component. We recommend that the ultimate deployment of this architecture only be made on platforms that support our virtualization-based trusted host component.

Send commands via 3G. An attacker may try to subvert the trusted host component when it is connected to the Internet via 3G/4G. For example, by sending commands to the trusted component, an attacker might attempt to remove their app from the blacklist or disable other benign apps for malicious intent. This involves the attacker sending commands to the special port that the trusted component on the device listens to. Apart from the fact that the attacker does not possess the keys for communicating with the device, the GGSN can be configured to block all communication to this port from other devices on the Internet as well as devices within the mobility network itself. The only communication that happens on this secure port is between the Mitigation Engine and the device.

Parasitic Malware. In some cases, malware can inject itself into other benign system processes. This is an especially popular technique used with malware that infects PCs running the Windows operating system [50]. In such cases, the trusted host component will identify the infected system process as malicious and kill it. While it is not clear if malware on mobile devices are following a similar trend, the host component can be trivially extended to identify the correct malware program by using techniques discussed in [50].

5.2 Scalability

When deployed in a real network, Triton should be able to handle millions of devices and traffic sent by them. While Triton is a prototype, it can leverage several well-known techniques to handle scale in a real network. The packet inspector can be a passive sniffer that can sniff traffic at very high speeds [25–26, 38]. Databases can handle queries and very fast lookups of terabytes of data [30, 22]. The workload of other components, such as the mitigation engine, the filter and the malware detectors can be split by using large clusters of machines and load balancing techniques already used in large data environments.

5.3 WiFi Offload

Due to both data cap and bandwidth limitations, mobile device users typically offload traffic to WiFi hotspots when available. Though Triton's network component does not have visibility into traffic that is offloaded via WiFi, each device still logs its own network activity in the host component. Since there are still millions of other customers currently connected to the mobility network at any given time, Triton's network detection algorithms maintain their global visibility and effectiveness. If a significant threat is detected, Triton can proactively communicate with devices that are currently connected to WiFi via an out of band channel, such as SMS or WAP push, to ensure that the device is not effected by the threat.

5.4 Limitations

The approach that we propose offloads detection to the network, while the host component does much of the mitigation and prevention. This is well suited to attacks that manifest themselves in the network, which the network provider has a good view of, such as, botnet command and control patterns, spread of worms, DDoS activities, malicious applications phoning home and so on. These activities are generally characteristic to mobile malware where attack activity is driven by economic incentives. However, malware can be designed to be largely focused on the device and operate without a network footprint. Our approach cannot detect malware that is purely device centric. In such cases, we assume the presence of other anti-virus software on the device that can detect such malware.

With network level detection techniques catching up, malware can become more stealthy and piggyback on genuine network traffic for communication with its controlling servers or spread to other devices. Stealthy malware is a challenge to detect in the network and would be challenging for our approach as well.

6 Related Work

We can classify related work in three main categories: network-based detec tion techniques and host based detection techniques that we leverage, and work that uses a combination of network and host based techniques.

There is a large body of work on network-based anomaly detection. Recent work has focused on detecting botnet command and control communication patterns based on protocols and heuristics [40], presence of a botnet infection cycle visible from the network [32], spatio-temporal correlations in network traffic [33] and clustering of network traffic [31] for detection of botnets. A similar body of work exists for other types of malware such as worms [52–53, 41], exploit code inside network flows [23], etc. Triton relies on efficient network-based detection techniques and can leverage them to improve its detection accuracy and efficacy.

We leverage techniques for securely inspecting the state of a user VM from a privileged VM as discussed in prior work [29, 18, 48]. Many VMM- based techniques have been proposed to detect malware running inside the user VM [43, 37]. Triton uses similar inspection techniques to find a malicious application but relies on network-based detectors to flag malicious traffic.

Zeng et al. developed a technique to improve the accuracy of botnet detection by using additional information on PCs, such as CPU and memory, via a component installed on the PC [54]. Srivastava et al. used a detector in the network to identify anomalous traffic, while a host component attributed this traffic to identify parasitic malware on the end device [50]. Our work differs in several ways from the above two works. We focus primarily on mobile devices as opposed to devices on the wireline network. We propose a defense in-depth approach that can detect, prevent and contain malware as opposed to only detecting it. The Triton architecture is designed to be integrated in a real mobility network and uses the on-host component only to prevent or mitigate an infection and therefore operates with a very low overhead on the mobile device.

Airmid [45] proposes a similar idea of remote repair where a network and host component act in tandem to recover from mobile malware infections. At the high level, our work shares the same insight but differs in the following ways: (a) We have a complete end to end working instance on a real 3G UMTS network where we address the design and implementation challenges encountered in realizing the architecture within the carrier. Airmid simulates the carrier side and assumes the existence of a server that can handle malicious traffic originating from mobile devices. (b) We report performance overheads on mobile devices supporting two different architectures - an inkernel component on HTC Aria phones and VMM based architecture running Xen on Viliv S5 devices. Airmid on the other hand only considers an in-kernel implementation. (c) We perform experiments with real mobile malware sam ples that have posed real threats on the Android market and show that our architecture can automatically identify malware and recover the end devices from malware infections. Airmid on the other hand uses prototype malware samples. Finally, Triton is built to interoperate with existing carrier equipment and does not require any changes within the carrier network elements, and therefore lends itself to easy adoption by carriers.

7 Conclusions

In this paper, we have described Triton, a new architecture for detection and containment of mobile malware. Triton employs a defense-in-depth approach where in-the-network malware detectors identify and prevent the spread of malware and communicate the threat to an on-the-phone trusted software component to identify and neutralize malware on the device. This allows mobility service providers to quickly respond to ongoing malware threats and contain malware on mobile devices, even in the absence of an anti-virus signature.

We reported on our experience and learnings from design and implementation of Triton in our 3G wireless lab for its server-side infrastructure components, and two prototype implementations for its client-side components namely, a kernel-level implementation on Android smart-phones, and a

VMM-based implementation on Linux Viliv devices. The prototype implementation of Triton successfully achieved infection prevention, effective detection, immediate containment, and fine-grained response on a diverse, representative set of real Android and Linux malware with a very low performance overhead. While our research described in this paper suggests that Triton has the potential to mitigate a large majority of mobile malware infections, we also discussed the potential counter attacks, scalability, and limitations of the Triton architecture.

Acknowledgements

We thank Tufan Demir, Gerry Eisenhaur and Mike Gagnon for their help with mobile malware analysis. We also thank Gustavo de los Reyes, Alex Bobotek and Dan Boneh for their insightful comments and feedback on the paper.

References

[1] Android and security. http://googlemobile.blogspot.com/2012/02/android-and-security.html.
[2] The bro network security monitor. http://bro-ids.org/.
[3] Cloudmark mobile platform. http://www.cloudmark.com/en/products/cloudmark-mobile-platform/how-it-works.
[4] Cyanogenmod wiki. http://wiki.cyanogenmod.com/index.php.
[5] Exercising our remote application removal feature. http://android-developers.blogspot.com/2010/06/exercising-our-remote-application.html.
[6] F-secure mobile threat report q2 2012.http://www.f-secure.com/weblog/archives/MobileThreatReport_Q2_2012.pdf.
[7] Infected apps in the android market. http://goo.gl/cFNiX.
[8] iphone security bug lets innocent-looking apps go bad. http://goo.gl/MMA2b.
[9] Lookout unveils the mobile threat network; verizon the first to adopt lookout api. http://goo.gl/bRxLt.
[10] New stealthy android spyware – plankton – found in official android market. http://www.csc.ncsu.edu/faculty/jiang/Plankton/.
[11] Okl4 microvisor.http://www.ok-labs.com/products/okl4-microvisor.
[12] Openggsn. http://sourceforge.net/projects/ggsn/.
[13] Remote kill and install on google android. http://jon.oberheide.org/blog/2010/06/25/remote-kill-and-install-on-google-android/.

[14] Security alert: Spamsoldier. https://blog.lookout.com/blog/2012/12/17 /security-alert-spamsoldier/.

[15] Snort. http://www.snort.org/.

[16] Universal mobile telecommunications system. http://en.wikipedia.org /wiki/Universal_Mobile_Telecommunications_System.

[17] Vmware mobile virtualization platform. http://www.vmware.com /products/mobile/overview.html.

[18] A. Baliga, L. Iftode, and X. Chen. Automated Containment of Rootkit Attacks. In *Elsevier, Computers & Security*, volume 27, 2008.

[19] F. Beck and O. Festor. Syscall Interception in Xen Hypervisor. In *MADYNES Technical Report*, 2009.

[20] J. Bickford, H. Lagar-Cavilla, A. Varshavsky, V. Ganapthy, and L. Iftode. Security versus Energy Tradeoffs in Host-Based Mobile Malware Detection. In *Proc. 9th Conference on Mobile Systems, Applications and Services*, 2011.

[21] Think Broadband. Download test files. http://www.thinkbroadband.com /download.html.

[22] F. Chang, J. Dean, S. Ghemawat, W. C. Hsieh, D. A. Wallach, M. Burrows, T. Chandra, A. Fikes, and R. E. Gruber. Bigtable: A Distributed Storage System for Structured Data. In *Proc. 7th USENIX Symposium on Operating Systems Design and Implementation*, 2006.

[23] R. Chinchani and E. van den Berg. A Fast Static Analysis Approach to Detect Exploit Code Inside Network Flows. In *Proc. 8th Symposium on Recent Advances in Intrusion Detection*, 2005.

[24] M. Christodorescu. Behavior-based Malware Detection. In *University of Wisconsin-Madison*, August 2007.

[25] C. Cranor, Y. Gao, T. Johnson, V. Shkapenyuk, and O. Spatscheck. Gigascope: High Performance Network Monitoring with an SQL Interface. In *Proc. Conference on Management of Data*, 2002.

[26] C. Cranor, T. Johnson, O. Spataschek, and V. Shkapenyuk. Gigascope: A Stream Database for Network Applications. In *Proc. Conference on Management of Data*, 2003.

[27] B. Dragovic, K. Fraser, S. Hand, T. Harris, A. Ho, I. Pratt, A. Warfield, P. Barham, and R. Neugebauer. Xen and the art of virtualization. In *Proc. 19th ACM Symposium on Operating Systems Principles*, 2003.

[28] W. Enck, P. Traynor, P. McDaniel, and T. La Porta. Exploiting Open Functionality in SMS-capable Cellular Networks. In *Proc. 12th Conference on Computer and Communications Security*, 2005.

[29] T. Garfinkel and M. Rosenblum. A Virtual Machine Introspection Based Architecture for Intrusion Detection. In *Proc. 10th Network and Distributed Systems Security Symposium*, 2003.

[30] R. Greer. Daytona and the Fourth-Generation Language Cymbal. In *Proc. Conference on Management of Data*, 1999.

[31] G. Gu, R. Perdisci, J. Zhang, and W. Lee. BotMiner: Clustering Analysis of Network Traffic for Protocol- and Structure-Independent Botnet Detection. In *Proc. 17th USENIX Security Symposium*, 2008.

[32] G. Gu, P. Porras, V. Yegneswaran, M. Fong, and W. Lee. BotHunter: Detecting Malware Infection Through IDS-Driven Dialog Correlation. In *Proc. 16th USENIX Security Symposium*, 2007.

[33] G. Gu, J. Zhang, and W. Lee. BotSniffer: Detecting Botnet Command and Control Channels in Network Traffic. In *Proc. 15th Network and Distributed System Security Symposium*, 2008.

[34] K. W. Hamlen, V. Mohan, M. M. Masud, L. Khan, and B. Thuraisingham. Exploiting an Antivirus Interface. *Computer Standards and Interfaces*, November 2009.

[35] J. Hwang, S. Suh, S. Heo, C. Park, J. Ryu, S. Park, and C. Kim. Xen on ARM: System Virtualization using Xen Hypervisor for ARM-based Secure Mobile Phones. In *Proc. 5th IEEE Consumer Communications and Networking Conference*, 2008.

[36] X. Jiang. An evaluation of the application verification service in android 4.2. http://www.csc.ncsu.edu/faculty/jiang/appverify/.

[37] X. Jiang, X. Wang, and D. Xu. Stealthy Malware Detection Through VMM-based "Out- of-the-Box" Semantic View Reconstruction. In *Proc. 14th Conference on Computers and Communications Security*, 2007.

[38] T. Johnson, S. M. Muthukrishnan, V. Shkapenyuk, and O. Spatscheck. Query-aware Partitioning for Monitoring Massive Network Data Streams. In *Proc. Conference on Management of Data*, 2008.

[39] S. T. Jones, A. C. Arpaci-Dusseau, and R. H. Arpaci-Dusseau. Antfarm: Tracking Processes in a Virtual Machine Environment. In *In Proc. USENIX Annual Technical Conference*, 2006.

[40] A. Karasaridis, B. Rexroad, and D. Hoeflin. Wide-scale Botnet Detection and Characterization. In *Proc. 1st Workshop on Hot Topics in Understanding Botnets*, 2007.

[41] H. Kim and B. Karp. Autograph: Toward Automated, Distributed Worm Signature Detection. In *Proc. 13th USENIX Security Symposium*, 2004.

[42] P. P. C. Lee, T. Bu, and T. Woo. On the Detection of Signaling DoS Attacks on 3G Wireless Networks. In *Proc. 26th IEEE Conference on Computer Communications*, 2007.

[43] L. Litty, H. A. Lagar-Cavilla, and D. Lie. Hypervisor Support for Identifying Covertly Executing Binaries. In *Proc. 17th USENIX Security Symposium*, 2008.

[44] L. W. McVoy and C. Staelin. lmbench: Portable Tools for Performance Analysis. In *Proc. USENIX Annual Technical Conference*, 1996.

[45] Y. Nadji, J. Giffin, and P. Traynor. Automated Remote Repair for Mobile Malware. In *Proc. 27th Annual Computer Security Applications Conference*, 2011.

[46] J. Oberheide and F. Jahanian. When Mobile is Harder Than Fixed (and Vice Versa): Demystifying Security Challenges in Mobile Environments. In *Proc. 11th Workshop on Mobile Computing Systems and Applications*, 2010.

[47] J. Oberheide and C. Miller. Dissecting Android's Bouncer. In *SummerCon*, 2012.

[48] B. D. Payne and W. Lee. Secure and Flexible Monitoring of Virtual Machines. In *Proc. 23rd Annual Computer Security Applications Conference*, 2007.

[49] P. Porras, H. Sadi, and V. Yegneswaran. An Analysis of the iKee.B iPhone Botnet. In *Security and Privacy in Mobile Information and Communication Systems*. 2010.

[50] A. Srivastava and J. T. Giffin. Automatic Discovery of Parasitic Malware. In *Proc. 13th Conference on Recent Advances in Intrusion Detection*, 2010.

[51] P. Traynor, M. Lin, M. Ongtang, V. Rao, T. Jaeger, P. McDaniel, and T. La Porta. On Cellular Botnets: Measuring the Impact of Malicious Devices on a Cellular Network Core. In *Proc. 16th Conference on Computer and Communications Security*, 2009.

[52] K. Wang, G. F. Cretu, and S. J. Stolfo. Anomalous Payload-Based Worm Detection and Signature Generation. In *Proc. 8th Symposium on Recent Advances in Intrusion Detection*, 2005.

[53] D. Whyte, E. Kranakis, and P. C. van Oorschot. DNS-based Detection of Scanning Worms in an Enterprise Network. In *Proc. 12th Network and Distributed System Security Symposium*, 2005.

[54] Y. Zeng, X. Hu, and K. G. Shin. Detection of Botnets Using Combined Host-and Network-level Information. In *Proc. 40th Conference on Dependable Systems and Networks*, 2010.

Biographies

Arati Baliga is an Independent Security Consultant and Adjunct Professor at the Department of Computer Science and Engineering at NYU Polytechnic School of Engineering. Her expertise and interests are in mobile, systems and network security. Previously, she was a Security Researcher at the AT&T Security Research Center where she worked on building carrier based defenses against mobile malware.

Arati obtained her Ph.D from the Department of Computer Science at Rutgers University. Her dissertation work on stealth malware detection won the outstanding paper award at the Annual Computer Security and Applications Conference (ACSAC 2008).

Jeffrey Bickford is a member of the AT&T Security Research Center. He is interested in mobile device security with a focus on using virtualization techniques to create a secure and robust mobile platform. His recent work has focused on designing a secure enterprise architecture for protection against APT attacks. Jeff also enjoys investigating real world attacks and analyzing mobile malware samples. He completed his M.S. at the Rutgers University Department of Computer Science.

Neil Daswani is a member of the information security group at Twitter. He was formerly the CTO of Dasient, Inc. prior to its acquisition by Twitter. Before co-founding Dasient, Neil had served in a variety of research, development, teaching, and managerial roles at Google, Stanford University, DoCoMo USA Labs, Yodlee, and Bellcore (now Telcordia Technologies). While at Stanford, Neil co-founded the Stanford Center for Professional Development Software Security Certification Program.

His areas of expertise include security, wireless data technology, and peer-to-peer systems. He has published extensively in these areas, frequently gives talks at industry and academic conferences, and has been granted several U.S. patents. He received a Ph.D. and a master's in computer science from Stanford University, and earned a bachelor's in computer science with honors with distinction from Columbia University.

Reinforcement Learning for Reactive Jamming Mitigation

Marc Lichtman and Jeffrey H. Reed

Wireless @ Virginia Tech, Virginia Tech, Blacksburg, VA, USA;
e-mail: {marcll, reedjh}@vt.edu

Received 25 March 2014; Accepted 25 March 2014;
Publication 2 July 2014

Abstract

In this paper, we propose a strategy to avoid or mitigate reactive forms of jamming using a reinforcement learning approach. The mitigation strategy focuses on finding an effective channel hopping and idling pattern to maximize link throughput. Thus, the strategy is well-suited for frequency-hopping spread spectrum systems, and best performs in tandem with a channel selection algorithm. By using a learning approach, there is no need to pre-program a radio with specific anti-jam strategies and the problem of having to classify jammers is avoided. Instead the specific anti-jam strategy is learned in real time and in the presence of the jammer.

Keywords: Reactive jamming, reinforcement learning, Markov decision process, repeater jamming, Q-learning.

1 Introduction

Wireless communication systems are becoming more prevalent because of their affordability and ease of deployment. Unfortunately, all wireless communications are susceptible to jamming. Jamming attacks can degrade communications and even cause total denial of service to multiple users of a system. As communications technology becomes more sophisticated, so does the sophistication of jammers. Even though jamming techniques such as

Journal of Cyber Security, Vol. 3 No. 2, 213–230.
doi: 10.13052/jcsm2245-1439.325

repeater jamming have been known for decades [2], recently there has been research into other forms of complex jamming with receiving and processing capabilities (reactive jamming) [8].

In this paper, we propose a strategy to mitigate or even avoid these reactive forms of jamming using a reinforcement learning (RL) approach. Through a learning approach, the problem of having to detect and classify which type of jammer is present in real time is avoided. In addition, there is no need to preprogram a radio with specific mitigation strategies; instead the strategy is learned in real time and in the presence of the jammer. The proposed mitigation strategy focuses on finding an effective channel hopping and idling pattern to maximize link throughput. Not only can this approach enable communications, which would otherwise fail in the presence of a sophisticated and reactive jammer, it can also act as an optimization routine that controls the link layer behavior of the radio.

The proposed strategy is well suited for a frequency-hopping spread spectrum (FHSS) system, which are widely used in modern wireless communications. The strategy could also be applied to an orthogonal frequency-division multiple access (OFDMA) system in which users hop between different subcarriers or groups of subcarriers. Countless users and systems depend on wireless communications and therefore it is important to secure them against jamming. While there exists many methods to counter barrage jamming (the most basic form of jamming), there are few methods that are designed to address the more intelligent behaviors a jammer can exhibit.

2 Related Works

Wireless security threats are typically broken up into two categories: cyber-security and electronic warfare (i.e. jamming). Electronic warfare attacks target the PHY and/or MAC layer of a communication system, while cyber-security attacks are designed to exploit the higher layers. In this paper we are only concerned with jamming, and in particular jamming of an intelligent nature. A series of intelligent jamming attack models are introduced in [8], including the reactive jammer model. The authors propose a basic detection algorithm using statistics related to signal strength and packet delivery ratio. For an overview on electronic warfare and jamming, we refer the reader to [1].

A RL or Markov decision process (MDP) approach has been previously used in the wireless domain for channel assignment [9], general anti-jamming in wireless sensor networks [10], and jammer avoidance in cognitive radio networks [4, 7]. The authors of [9] apply reinforcement learning to the problem of channel assignment in heterogeneous multicast terrestrial communication systems. While this paper does not deal with jamming, it has similar concepts to the techniques proposed in this paper. The authors of [10] propose an anti-jamming scheme for wireless sensor networks. To address time-varying jamming conditions, the authors formulate the anti-jamming problem of the sensor network as a MDP. It is assumed that there are three possible anti-jamming techniques: transmission power adjustment, error-correcting code, and channel hopping. These techniques are not explored any further; the set of actions available to the radio is simply which technique is used. While this work is similar to the technique described in this paper, it greatly generalizes the anti-jamming strategies. In other words, this work does not offer a jamming strategy, it offers a method of choosing the best jamming strategy from a given set. The authors of [7] use a MDP approach to derive an optimal anti-jam strategy for secondary users in a cognitive radio network. For the jammer model, the authors use reactive jammers seeking to disrupt secondary users and avoid primary users. In terms of actions, in each timeslot the secondary user must decide whether to stay or hop frequencies. The authors propose an online learning strategy for estimating the number of jammers and the access pattern of primary users (this can be thought of as channel availability). Even though the authors use a reactive jammer model similar to the one described in this paper, they assume the jammer is always successful, and the entire analysis is within the context of dynamic spectrum access.

To the best of our knowledge, there have been no RL or MDP based approaches designed to mitigate a wide range of reactive jamming behaviors. This paper will provide initial insights into the feasibility and suitability of such an approach.

3 System Model and Problem Formulation

Consider the typical wireless communications link, with the addition of a jammer that both receives the friendly signal (but not necessarily demodulates it) and transmits a jamming signal, as shown in Figure 1. For the sake of simplicity we will only consider a unidirectional link, although this analysis also applies to bidirectional links, that may be unicast or broadcast, as well as a collection of links.

While reactive jamming can take on different forms, we will broadly define the term as any jammer that is capable of sensing the link and reacting to sensed information. We will assume this sensed information is in the form of

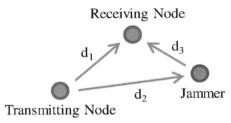

Figure 1   System model of a transmitter, receiver, and reactive jammer

the presence or absence of energy, because any additional information such as modulation scheme or actual data would be irrelevant for this mitigation strategy. A simple example of a reactive jammer is one that senses the spectrum for activity, and immediately transmits wideband noise when it senses any activity [8]. This strategy allows the jammer to remain idle while the channel is idle and thus save power and avoid being easily detected. Another form of reactive jamming, commonly known as repeater or follower jamming [6], works by immediately transmitting what it receives with noise added to it. This can be modeled as a jammer that senses a set of channels, and immediately transmits noise on any channel that appears to be active.

Reactive jamming is only feasible when the geometry of the system is such that the jammer's transmitted signal reaches the target receiver before it hops to a new channel or stops transmitting. As such, reactive jamming is only possible when the jammer is physically located near or between the target transmitter and receiver. If η represents the fraction of each hop duration that must remain not jammed for communications to succeed, then we have the following inequality limiting the distances d_2 and d_3 [6]

$$d_2 + d_3 \leq (\eta T_d - T_j)c + d_1 \tag{1}$$

where T_d is the hop duration, T_j is the jammer's processing time, c is the speed of light, and $d_1, d_2,$ and d_3 are the distances indicated in Figure 1. In addition to this limitation, the jammer-to-signal ratio at the receiving node must be high enough to degrade quality of service. In this paper we assume the jammer is close enough to the transmitter and receiver, and that the jammer-to-signal ratio is significantly high during periods of jamming.

As part of the analysis and simulation we will investigate two specific reactive jamming models. The first, labeled in this paper as simply "reactive jamming", will be defined as a jammer that successfully jams any transmission that remains active for too long, regardless of the channel/frequency in use.

The second jammer model is based on repeater jamming, and it is described as a jammer which successfully jams any transmission that remains on the same channel/frequency for too long. While there are other ways to formulate reactive jamming models, the analysis and simulation in this paper will focus on these two. More formal definitions of these two jammer models is as follows:

- **Reactive Jammer** - Begins jamming any transmission that remains active for more than N_{REACT} time steps, and will only cease jamming once the target is idle for at least N_{IDLE} time steps.
- **Repeater Jammer** - Begins jamming any transmission that remains on the same channel for more than N_{REP} time steps.

In this analysis we will investigate a transmitter and receiver pair that can hop among a certain number of channels using a FHSS approach, or any other approach that involves radios capable of changing channels. Therefore, at any time step, the transmitter has the option to either remain on the channel, change channel, or go idle. Because the actions of the transmitter must be shared with the receiver beforehand, it is expected that decisions are made in advanced.

It is assumed that channel quality indicators (e.g. whether or not the information was received) are sent back to the transmitter on a per-hop basis. These indicators could be binary (indicating an ACK or NACK), or they could take on a range of values indicating the link quality. Lastly, it is assumed that the receiver is not able to simply detect the presence of a jammer.

4 Strategy for Mitigation of Reactive Jamming

The mitigation (a.k.a. anti-jam) strategy described in this paper is based on modeling the system as a MDP, where the transmitter is the decision maker, and using RL to learn a strategy for dealing with the broad category of reactive jamming. This strategy will be in the form of a channel hopping pattern, where going idle is considered as hopping to the "idle channel" for a certain duration. However, we are not concerned with choosing the best channel to transmit on at any given time, nor identifying corrupt channels that have excessive noise. The mitigation strategy described in this paper is designed to work in tandem with this kind of algorithm, i.e. one that indicates which specific channels are suitable for use and which are not. Likewise, we are not concerned with the PHY-layer waveform characteristics that the transmitter or jammer uses (i.e.

bandwidth, modulation, type of noise, etc.). Adaptive modulation and coding can be performed alongside the proposed strategy.

4.1 Reinforcement Learning Background

RL is the subset of machine learning concerned with how an agent should take actions in an environment to maximize some notion of cumulative reward. The agent is the entity interacting with the environment and making decisions at each time interval, and in this paper we will consider the transmitter as the agent (although the actions it chooses must be forwarded to the receiver). An agent must be able to sense some aspect of the environment, and make decisions that affect the agent's state. For example, reinforcement learning can be used to teach a robot how to walk, without explicitly programming the walking action. The robot could be rewarded for achieving movement in a forward direction, and the robot's action at each time step could be defined as a set of angular servo motions. After trying a series of random motions, the robot will eventually learn that a certain pattern of motion leads to moving forward, and thus a high cumulative reward. In this paper, we apply this concept to a transmitter that learns how to hop/idle in a manner that allows successful communications under a sophisticated reactive jamming attack.

There are four main components of a RL system: a policy, reward, value function, and the model of the environment [5]. A policy (denoted as π) defines how the agent will behave at any given time, and the goal of a RL algorithm is to optimize the policy in order to maximize the cumulative reward. A policy should contain a stochastic component, so that the agent tries new actions (known as exploration). A reward, or reward function, maps the current state and action taken by the agent to a value, and is used to indicate when the agent performs desirably. In a communication system, a possible reward function may combine the throughput of a link, spectral efficiency, and power consumption. While the reward function indicates what is desirable in the immediate sense, the value function determines the long-term reward. A state may provide a low immediate reward, but if it leads to other states that provide a high reward, then it would have a high "value".

The model of the environment is used to either predict a reward that has not been experienced yet, or simply determine which actions are possible for a given state. For example, it is possible to create a RL agent that learns how to play chess, and the environment would be a model of the chess board, pieces, and set of legal moves.

In RL, the environment is typically formulated as a MDP, which is a way to model decision making in situations where outcomes are partially random and partially under the control of the decision maker. The probability of each possible next sate, s', given the current state s and action a taken, is given by [5]

$$P_{ss'}^a = Pr\{s_{t+1} = s'|s_t = s, \ a_t = a\} \tag{2}$$

Equation 2 provides what are known as transition probabilities, and because they are only based on the current state and action taken, it assumes a memoryless system and therefore has the Markov property. The expected reward (obtained in the next time step) for a certain state-action pair is given by Equation 3. The goal of a learning agent is to estimate these transition probabilities and rewards, while performing actions in an environment.

$$R_{ss'}^a = E\{r_{t+1}|s_{t+1} = s', \ s_t = s, \ a_t = a\} \tag{3}$$

In order for an agent to take into account the long-term reward associated with each action in each state, it must be able to predict the expected long-term reward. For a certain policy π, we calculate the expected return from starting in state s and taking action a as [5]

$$Q^\pi(s, a) = E_\pi \left\{ \sum_{k=0}^{\infty} \gamma^k r_{t+k+1}|s_t = s, \ a_t = a \right\} \tag{4}$$

where γ is known as the discount rate, and represents how strongly future rewards will be taken into account. Equation 4 is known as the action-value function, and in a method known as Q-Learning, the action-value function is estimated based on observations. While performing actions in an environment, the learner updates its estimate of $Q(s_t, a_t)$ as follows [5]:

$$Q(s_t, a_t) \leftarrow Q(s_t, a_t) + \alpha \left[r_{t+1} + \gamma \max_a Q(s_{t+1}, a) - Q(s_t, a_t) \right] \tag{5}$$

where r_{t+1} is the reward received from taking action a, and α is the learning rate, which determines how quickly old information is replaced with new information. Because Q-Learning is an iterative algorithm, it must be programmed with initial conditions ($Q(s_0, a_0)$). Optimistically high values are typically used for initialization, to promote exploration. However, even once some initial exploration is performed, there needs to be a mechanism that prevents the agent from simply sticking to the best policy at any given

actions apply to all jammer models. time. An approach known as Epsilon-greedy forces the agent to take the "best action" with probability $1-\epsilon$, and take a random action (using a uniform probability) with probability ϵ. Epsilon is usually set at a fairy high value (e.g. 0.95) so that a majority of the time the agent is using what it thinks is the best action. For an in-depth tutorial on MDPs and RL, we refer the reader to [5].

4.2 Markov Decision Process Formulation

We will now formulate a MDP used to model the transmitter's available states and actions. The states exist on a two dimensional grid, in which one axis corresponds to the time that the transmitter has been on a given channel (including the "idle channel"), and the other axis corresponds to the time the transmitter has been continuously transmitting. Time is discretized into time steps, and we will assume the step size is equal to the smallest period of time in which the transmitter must remain on the same channel. Figure 2 shows the state space and actions available in this MDP.

The transmitter will always start in the top-left state, which corresponds to being idle for one time step. It then must choose whether to remain idle, or "change channel" (which can be interpreted as "start transmitting" when coming from an idle state). If it decides to change channel, then in the next time step it must decide whether to remain on that channel or change to a new channel, which we will assume is chosen randomly from a list of candidate channels. It should be noted that the result of each action is deterministic,

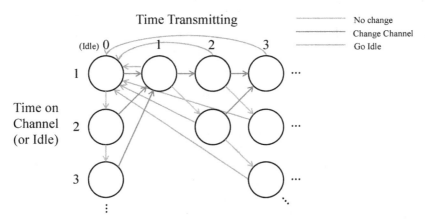

Figure 2   Markov Decision Process associated with hopping channels and going idle. These states and actions apply to all jammer models

however the rewards may contain a stochastic component. Due to what the states represent, the MDP is theoretically infinitely long in directions indicated by ellipsis in Figure 2. However, in practical systems the width and height of the MDP must be limited, as we discuss later.

The reward associated with each state transition is based on the actual data throughput that occurs during the time step. As such, the rewards are based on the jammer model, and may be stochastic in nature. Figure 3 shows the rewards associated with a transmitter and receiver operating in the presence of a reactive jammer with $N_{REACT} = 3$ and $N_{IDLE} = 1$ (model and parameters defined in the previous section). This example shows that when the radio is transmitting for more than three continuous time steps, the link becomes jammed (red states) and the reward becomes zero until the jammer goes idle and then starts transmitting again (the radio is not rewarded while idle). Although the rewards are shown on top of each state, they are actually associated with the previous state and action taken, and won't always be equal for a given resulting state. The numbers 1, 1.3, and 1.47 are examples to demonstrate the fact that remaining on the same channel is more favorable than switching channels, due to the time it takes to switch frequencies. In a

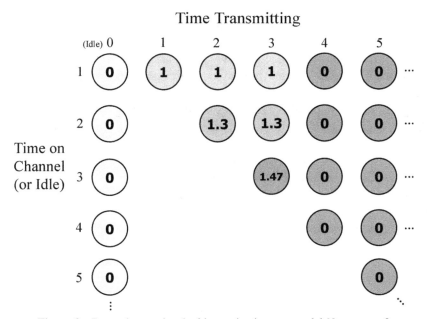

Figure 3 Rewards associated with reactive jammer model $N_{REACT} = 3$

Table 1   Summary of how to cast this mitigation approach into a RL framework

Environment States exist on a two dimensional grid, in which one axis corresponds to the time the transmitter has been on a given channel (including the "idle channel"), and the other axis corresponds to the time the transmitter has been continuously transmitting.
Agent's Actions are to either 1) remain idle or 2) "change channel" which can be interpreted as "start transmitting" when coming from an idle state.
State Transition Rules are deterministic (although a stochastic component due to external factors could be added) and based on the action taken.
Reward Function is a value proportional to the data throughput that occurred during the time step (not known until feedback is received).
Agent's Observations include the state it is currently in, and the reward achieved from each state-action pair.
Exploration vs. Exploitation is achieved using the Epsilon-greedy approach, in which the agent chooses a uniformly random action a small fraction of the time.
Task type is continuing by nature, but could be treated as episodic where each episode is an attempt to transmit for N time steps.

real implementation the reward would be based on the achieved throughput or quality of the link; not a model. A summary of how to cast this mitigation approach into a RL framework is given in Table 1.

Now that the states, actions, and rewards are established, we can investigate the learning process of the transmitter in the presence of various types of reactive jammers. In RL, the agent (in this case, the transmitter) learns by trying actions and building statistics associated with each state-action pair. At the beginning of the learning process, the agent has no information about the environment, and must try random actions in any state. After a period of learning the agent eventually chooses what it thinks is the best action for each state in terms of the predicted long-term reward. The Epsilon-greedy approach forces the agent to never consider itself "finished" with learning.

Under a reactive jammer with a certain N_{REACT} and when $N_{IDLE} = 1$, the optimal policy is to remain on the same channel for N_{REACT} time steps, and then go idle for one time step. Three optimal policies are shown in Figure 4, correspond to $N_{REACT} = 1, 2$, and 3. Each optimal policy resembles a loop that starts at idle for one time step and proceeds to transmit on the same channel for N_{REACT} time steps. In a real-world scenario, it takes the transmitter many

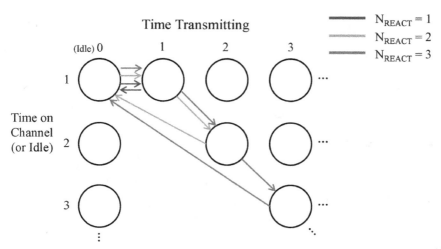

Figure 4 Optimal policies in the presence of three different reactive jammers

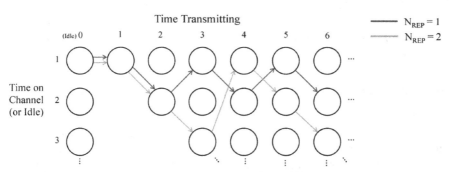

Figure 5 Optimal policies in the presence of two different repeater jammers

time steps to establish that this is the best policy to take, because it must explore each state-action multiple times to build reliable statistics.

The optimal policy for a repeater jammer is shown in Figure 5, using $N_{REP} = 1$ and 2. This zigzag pattern indicates constant-rate frequency hopping, which is well-established as the typical method for avoiding repeater jamming [6]. Unfortunately the optimal policy will always be infinitely long in the horizontal direction. To take this into account, the learning process can involve resetting the current state to the top-left state after a certain period of time continuously transmitting. This will have minimal influence on learning the optimal policy as long as the state space spans enough time steps to take into account the slowest (i.e. the highest value of N_{REP}) perceivable repeater jammer.

Using the approach described in this paper, there is no need to perform "jammer classification". As such, the mitigation strategy will remain effective over a wide range of jamming behaviors, and may even be able to deal with new jamming behaviors that were not considered during development.

4.3 Knowledge Decay

The last component of the proposed mitigation strategy is taking into account a changing environment. A given jammer may only be present for a short period of time, and link performance would degrade if the transmitter were sticking to its acquired knowledge. As such, the learning engine must incorporate some form of knowledge decay. Due to the nature of Q-Learning, the learning rate α can be used as a form of knowledge decay, by setting it low enough so that the learner can react to a changing environment. A proper value for α would be based on how quick the transmitter is able to learn optimal policies for a variety of jammers. A detailed investigation on approaches of knowledge decay/forgetting is beyond the scope of this paper, but for more information we refer the reader to [5].

4.4 Comparison with Traditional Parameter Optimization

Finding an effective channel hopping and idling pattern in the presence of a reactive jammer could also be performed by optimizing the hopping rate and transmission duty cycle. This can also be thought of as adjusting T_{ON} and T_{OFF}; the transmission and idle time of a transmitter, assuming it hops frequencies after each transmission. This type of approach is often used in cognitive radio [3]. If $T_{OFF} = 0$, then T_{ON} becomes the hopping rate. Any number of optimization approaches could be used to tune these two parameters. However, even though this simpler approach can take into account the two specific jammer models described in this paper, it does not have the flexibility inherent to the RL approach. For example, consider the scenario involving a reactive jammer with $N_{REACT} = 4$, $N_{IDLE} = 1$ and a repeater jammer with $N_{REP} = 1$, both targeting the friendly node simultaneously. The optimal transmission strategy would be to hop channels every time step, but also go idle for one time step after the fourth consecutive transmission (a strategy which is likely not possible with traditional parameter optimization). In addition, if the actual jammer behavior experienced by the transmitter does not match any models developed during creation of the mitigation strategy, then added flexibility may mean the difference between communicating and being fully jammed.

5 Simulation Results

In this section, we present some simulation results to show proof of concept of our proposed technique. To simulate this RL based mitigation strategy, a link layer simulation framework was created, which included the jammer models described in this paper. Q-learning was chosen as the RL technique [5]. In terms of Q-learning parameters, a learning rate, α, of 0.95 (the transmitter will quickly use learned knowledge) and discount factor, γ, of 0.8 was used for the simulations. This relatively low discount factor was used because of the cyclic nature of the optimal policies. Figure 6 shows the reward over time for various jamming models, depicting the learning process until saturating to an effective policy with constant reward. Because the reward from each time step is proportional to link throughput, the results can be interpreted as throughput over time. The barrage jammer was modeled by causing jamming with 20% probability at each time step, regardless of how long the transmitter has been transmitting or on a given channel. This can be thought of as a nonreactive jammer that is always transmitting, but at a jammer-to-signal ratio that is not quite high enough to cause complete denial of service.

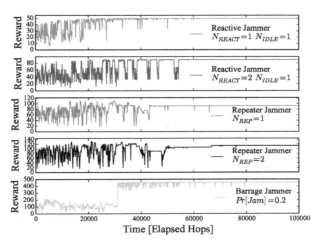

Figure 6 Simulation results showing the learning process over time in the presence of different jammers

The maximum achievable reward under each jamming behavior varies, which is expected (e.g. $N_{REACT} = 2$ will allow using a higher duty cycle than $N_{REACT} = 1$). Although not depicted in Figure 6, it should be noted that the transmitter learned the optimal policy (discussed in the previous section) only during the reactive jamming and barrage jamming scenarios. In both repeater jamming scenarios the learned policy did not traverse the entire zigzag pattern on the MDP, which is the optimal policy for the repeater jamming model as discussed earlier. Rather, the transmitter would go idle on occasion, which would essentially reset the zigzag pattern. Hence, the reward achieved under repeater jamming was not the maximum possible reward. Under barrage jamming the optimal policy for the transmitter would be to remain transmitting on the same channel indefinitely, which occurred after around 50,000 time steps, except for the occasional channel hop (as indicated by the small dips in the plot). This demonstrates how the proposed strategy can work under non-reactive jamming, despite not being designed to do so, and even provide better throughput than a constant-rate FHSS approach by avoiding the overhead associated with changing channels.

It should be noted that the time taken to learn an effective strategy for a given jammer is a function of the learning rate parameter and learning technique (Q-learning in this case). Results in Figure 6 show a learning time between 30,000 and 50,000 time steps, which is one or two seconds in a system where the minimum hop duration is on the order of tens of microseconds. While this may seem long compared to a radio that is preprogrammed with specific anti-jam strategies, it is unlikely that the presence of different jammers will change within seconds. In addition, the preprogrammed radio must spend time classifying the type of jammer present in order to know which mitigation scheme to use; a process which is not needed for the proposed strategy. We remind the reader that although wireless channel conditions are known for changing within milliseconds, the proposed strategy is meant to counter the adversary; not traditional channel imperfections such as fading or doppler shift.

6 Conclusions

In this paper, we have developed a RL based strategy that a communication system can use to deal with reactive jammers of varying behavior by learning an effective channel hopping and idling pattern. Simulation results provide a proof of concept and show that a high-reward strategy can be established within a reasonable period of time (the exact time being dependent on the duration of a time step).

This approach can deal with a wide range of jamming behaviors, not known a priori. Without needing to be preprogrammed with anti-jam strategies for a list of jammers, our approach is able to better adapt to the environment. The proposed technique is best used in tandem with an algorithm that finds a favorable subset of channels to use, as well as modern optimization techniques such as adaptive modulation and forward error correction. In future work we will investigate expanding the MDP state space to take into account additional factors, as well as explore more stochastic jammer models. In addition, it is likely that the RL procedure can be tuned to provide even greater performance.

Acknowledgement

This material is based on research sponsored by the Air Force Research Laboratory under agreement number FA9453-13-1-0237. The U.S. Government is authorized to reproduce and distribute reprints for Governmental purposes notwithstanding any copyright notation thereon. The views and conclusions contained herein are those of the authors and should not be interpreted as necessarily representing the official policies or endorsements, either expressed or implied, of the Air Force Research Laboratory or the U.S. Government.

Refernces

[1] David L Adamy. *EW 101*. Artech House, 2001.
[2] Joseph Aubrey Boyd, Donald B Harris, Donald D King, and HW Welch Jr. Electronic countermeasures. *Electronic Countermeasures*, 1, 1978.
[3] S.M. Dudley, W.C. Headley, M. Lichtman, E.Y. Imana, Xiaofu Ma, M. Abdelbar, A. Padaki, A. Ullah, M.M. Sohul, Taeyoung Yang, and J.H. Reed. Practical issues for spectrum management with cognitive radios. *Proceedings of the IEEE*, 102(3): 242–264, March 2014.
[4] Shabnam Sodagari and T Charles Clancy. An anti-jamming strategy for channel access in cognitive radio networks. In *Decision and Game Theory for Security*, pages 34–43. Springer, 2011.
[5] Richard S Sutton and Andrew G Barto. *Reinforcement learning: An introduction*, volume 1. Cambridge Univ Press, 1998.

[6] Don J Torrieri. Fundamental limitations on repeater jamming of frequency-hopping communications. *Selected Areas in Communications, IEEE Journal on*, 7(4): 569–575,

[7] Yongle Wu, Beibei Wang, and KJ Ray Liu. Optimal defense against jamming attacks in cognitive radio networks using the markov decision process approach. In *IEEE Global Telecommunications Conference 2010*, pages 1–5. IEEE, 2010.

[8] Wenyuan Xu, Wade Trappe, Yanyong Zhang, and Timothy Wood. The feasibility of launching and detecting jamming attacks in wireless networks. In *Proceedings of the 6th ACM international symposium on Mobile ad hoc networking and computing*, pages 46–57. ACM, 2005.

[9] Mengfei Yang and David Grace. Cognitive radio with reinforcement learning applied to heterogeneous multicast terrestrial communication systems. In *Cognitive Radio Oriented Wireless Networks and Communications, 2009*, pages 1–6. IEEE, 2009.

[10] Yanmin Zhu, Xiangpeng Li, and Bo Li. Optimal adaptive antijamming in wireless sensor networks. *International Journal of Distributed Sensor Networks*, 2012, 2012.

Biographies

Marc Lichtman is a Ph.D. student at Virginia Tech under the advisement of Dr. Jeffrey H. Reed. His research is focused on designing anti-jam approaches against sophisticated jammers, using machine learning techniques. He is also interested in analyzing the vulnerability of LTE to jamming. Mr. Lichtman received his B.S. and M.S. in Electrical Engineering at Virginia Tech in 2011 and 2012 respectively.

Jeffrey H. Reed currently serves as Director of Wireless @ Virginia Tech. He is the Founding Faculty member of the Ted and Karyn Hume Center for National Security and Technology and served as its interim Director when founded in 2010. His book, Software Radio: A Modern Approach to Radio Design was published by Prentice Hall. He is co-founder of Cognitive Radio Technologies (CRT), a company commercializing of the cognitive radio technologies; Allied Communications, a company developing technologies for embedded systems. In 2005, Dr. Reed became Fellow to the IEEE for contributions to software radio and communications signal processing and for leadership in engineering education. He is also a Distinguished Lecture for the

IEEE Vehicular Technology Society. In 2013 he was awarded the International Achievement Award by the Wireless Innovations Forum. In 2012 he served on the Presidents Council of Advisors of Science and Technology Working Group that examine ways to transition federal spectrum to allow commercial use and improve economic activity.

www.ingramcontent.com/pod-product-compliance
Lightning Source LLC
LaVergne TN
LVHW012331060326
832902LV00011B/1825